Environmental Oceanography

Topics and Analysis

Daniel C. Abel

Senior Fellow
U.S. Partnership for Education for Sustainable Development
Director, CCU Campus and Community Sustainability Initiative
and
Department of Marine Science
Coastal Carolina University
Conway, SC

Robert L. McConnell

Senior Fellow
U.S. Partnership for Education for Sustainable Development
Emeritus Professor
Environmental Science and Geology
University of Mary Washington
Fredericksburg, VA

JONES AND BARTLETT PUBLISHERS

Sudbury, Massachusetts

BOSTON   TORONTO   LONDON   SINGAPORE

World Headquarters

Jones and Bartlett Publishers
40 Tall Pine Drive
Sudbury, MA 01776
978-443-5000
info@jbpub.com
www.jbpub.com

Jones and Bartlett Publishers Canada
6339 Ormindale Way
Mississauga, Ontario L5V 1J2
Canada

Jones and Bartlett Publishers
International
Barb House, Barb Mews
London W6 7PA
United Kingdom

Jones and Bartlett's books and products are available through most bookstores and online booksellers. To contact Jones and Bartlett Publishers directly, call 800-832-0034, fax 978-443-8000, or visit our Web site, www.jbpub.com.

Substantial discounts on bulk quantities of Jones and Bartlett's publications are available to corporations, professional associations, and other qualified organizations. For details and specific discount information, contact the special sales department at Jones and Bartlett via the above contact information or send an email to specialsales@jbpub.com.

Production Credits

Chief Executive Officer: Clayton Jones
Chief Operating Officer: Don W. Jones, Jr.
President, Higher Education and Professional Publishing: Robert W. Holland, Jr.
V.P., Sales and Marketing: William J. Kane
V.P., Design and Production: Anne Spencer
V.P., Manufacturing and Inventory Control: Therese Connell
Publisher, Higher Education: Cathleen Sether
Acquisitions Editor, Science: Molly Steinbach
Managing Editor, Science: Dean W. DeChambeau
Associate Editor, Science: Megan R. Turner
Editorial Assistant, Science: Caroline Perry
Production Editor: Dan Stone
Senior Marketing Manager: Andrea DeFronzo
Composition: diacriTech, Chennai, India
Interior Design: Anne Spencer
Cover Design: Kate Ternullo
Cover Image: © Courtesy of Grant Johnson.
Assistant Photo Researcher: Jessica Elias
Printing and Binding: Malloy, Inc.
Cover Printing: Malloy, Inc.

Library of Congress Cataloging-in-Publication Data

Abel, Daniel C.
  Environmental oceanography: topics and analysis/Daniel C. Abel, Robert L. McConnell.
    p. cm.
  Includes bibliographical references and index.
  ISBN 978-0-7637-6379-4
  1. Oceanography. 2. Ocean–Environmental conditions. I. McConnell, Robert L. II. Title.
  GC28.A34 2009
  333.91'64–dc22
                                                                                            2008051338

6048
Printed in the United States of America
13 12 11 10 09   10 9 8 7 6 5 4 3 2 1

CONTENTS

To the Student

As environmental scientists, we care deeply about the health of the ocean, and we feel that you, as a responsible citizen who will have to make increasingly difficult choices in the years ahead, need to be concerned about it as well. We hope you find *Environmental Oceanography: Topics and Analysis* to be a provocative introduction to a number of salient marine issues, including many that you may have never thought about. Within the text are self-contained case studies with questions that ask you to think critically about each issue, in some cases using simple math. These are real-life issues, not hypothetical ones, and you need certain basic skills to understand them fully. They are as follows:

- You must be familiar with and be able to use the units of the metric system. (Do not worry, we provide a review in the Appendix.)

- You must be able to use a few simple mathematical formulas to quantify the issues that you are studying, and you must be able to carry out the calculations accurately. (We help you with these, too.)

- You must rigorously and continuously assess your thinking and apply certain critical thinking skills and techniques when discussing the implications of your calculations. (Critical thinking is so important that we have written an entire section on it.)

We understand that many students have some "math anxiety," or are a bit rusty in the use of some math skills; thus, we use a step-by-step method to take you through many of the calculations in the issues, and we have an appendix devoted to introducing or reviewing the math concepts we use in the book. Math proficiency is one of the skills necessary for fully understanding environmental issues; without these skills, your only option is to make choices on the basis of which "expert" you believe. Becoming educated, however, is much more than simply acquiring math skills. Therefore, we have two additional fundamental objectives: to provide you with the knowledge and intellectual standards necessary to apply critical thinking to marine environmental studies and to foster your ability to evaluate other issues critically.

To the Instructor

Environmental Oceanography: Topics and Analysis is an interactive textbook/case studies book that is designed to teach students to learn about pressing marine environmental issues using critical thinking and basic math. This book is appropriate for use at several different levels. Because we introduce most, if not all, of the essential topics, including introductory chapters on physical oceanography and marine ecology, it could be used as a stand-alone text, supplemented with web-based activities and readings, in an introductory or advanced course. The book could also be used in a laboratory or recitation setting. Additionally, the book could be used in the new generation of sustainability and sustainable development courses springing up on campuses worldwide. We have presented our ideas at educational forums, and educators from high school through college have enthusiastically endorsed our approach.

The text uses an innovative approach to teaching environmental oceanography, consisting of marine environmental issues presented as self-contained analytical exercises, with information and questions on sustainability integrated throughout the text.

The traditional approach to teaching oceanography at the introductory level includes a comprehensive survey of all topics within the broad discipline of ocean science. Ironically, it is the topic usually presented last—marine environmental issues—that grasps and holds the attention of most students and thus can be an effective vehicle for teaching fundamental concepts of marine science. It is also, in our opinion, the most important topic and unfortunately is the one most likely to be omitted or only superficially covered because of time constraints.

The text of *Environmental Oceanography: Topics and Analysis* begins with sections on the scientific method, principles of critical thinking, and logic. Next is an interactive exercise on the regulatory role of government, an introduction to sustainability and marine environmental issues, and an overview of essential principles of physical oceanography and marine ecology. These are followed by twenty-five interactive issues, which should take students between one and three hours each to complete. Each issue begins with a discussion of a pressing marine environmental issue, including defined terms, followed by a self-contained activity designed to develop students' critical thinking skills in a deliberate and structured way. By their nature, these issues require students to integrate topics from a range of subdisciplines to measure, analyze, and evaluate the issue, using the scientific method.

Most of these issues also have one or more media analyses associated with them. These media analyses are based primarily on short National Public Radio reports, which are reliably accessed and require only Windows Media Player, RealPlayer, iTunes, or a similar program, and Internet access. These media analyses usually represent different perspectives on the issue, and they allow students to analyze the spoken word critically, an important skill for students who receive most of their information in formats other than writing.

The topics we have selected for this book are those that we feel are among the most important marine issues facing society today. Our choices are also based on the comprehensive and authoritative 2003 Pew Oceans Commission Report on the state of the oceans.

The issues and critical thinking questions have been designed to be provocative. Some deal with protecting marine resources. Others address geohazards that coastal and island residents face. Some have to do with questions of values, as in Issue 15, which examines the impact of LFA SONAR on marine mammals. On one hand, these issues help students understand and appreciate the complexities of the global environment and the fact that many environmental issues are interrelated. On the other hand, they show how simple decisions may shape the success or failure of entire marine species or environments.

We stress the interdependency of human societies where the oceans are concerned. Because oceans are connected and pollution is free to move throughout ocean basins, all coastal nations are at risk, and no one country can solve all marine problems. In this context, we have included an introductory activity, "The Role of Government," so that students can assess their own position on the appropriate role of government and decide what level of leadership should address marine environmental issues.

One of our major objectives is to help develop math literacy (numeracy)—not necessarily arcane math, but the kind of arithmetic needed to properly quantify environmental issues. Such skills include the ability to manipulate large numbers using scientific notation and exponents and the ability to use growth equations. A lack of math skills often leaves students unprepared to deal with the complexity of today's environmental issues. We take pains to ensure that students of all majors and all levels of math anxiety are not intimidated by the calculations in the book.

The book is accompanied by an instructor's manual with an answer key which can be obtained through your Jones and Bartlett Publishers representative.

Finally, we encourage adopters to e-mail us with corrections, viewpoints, and especially ideas for new issues.

■ How to Use This Book Effectively

Here's how we use *Environmental Oceanography: Topics and Analysis* in our classes. First, we dedicate a class or a portion of a laboratory early in the semester to introducing students to the principles of critical thinking. In this period, we ask students to list characteristics of critical thinking (or higher order thinking, or just plain good, effective thinking). More often than not, the class' list encompasses many of the standards of critical thinking that are contained in the book's section on critical thinking. As we go over these characteristics, we emphasize clarity, accuracy, awareness of assumptions, and continuous self-assessment, as well as the importance of applying critical thinking to environmental (and other) issues. We then analyze passages from letters to the editor, newspaper commentary articles, and articles from popular magazines.

It is also worthwhile, before using this book, to confront students' math anxiety early and attempt to reassure them that they are capable of doing math, although they may be rusty and require some practice and assistance.

One may assign issues or blocks of issues as group projects. Students can collaborate using a web page, "Blackboard," or other form of electronic discussion group for your class.

We have found that many students who would be reluctant to participate in a classroom discussion will willingly contribute in the relative anonymity of an electronic discussion group. Students can copy their math and send it to others over the system, thereby checking each other's work. You can send comments to the groups if you feel they are on the wrong track and can commend them and encourage them if they are making progress. You can send them questions arising from their own discussions and can respond immediately to their inquiries. They can exchange information on the analysis questions, and you can discuss these questions with the students as their work evolves.

Regardless of whether you are teaching a large or small class, you can have students work in groups of two to four members. They can debate the issues and grade each other, or they can turn in their work in the normal fashion. Whatever the size of your class, work done on these issues can take the place of one or more exams, freeing you for other activities and providing students with a less threatening way to earn class credit than an all-exam format. We believe that students retain more information from work on projects and reports than from cramming for tests.

■ Acknowledgments

We thank Shoshanna Goldberg, Molly Steinbach, Daniel Stone, Jessica Elias, Andrea DeFronzo, and their staffs at Jones and Bartlett Publishers for their professionalism, skill, accessibility, receptiveness, organization, and not least, their enthusiasm for this project. Working with them was a genuine pleasure.

PART 1

Introduction to Science and the Use of Information

Science as a Way of Knowing

Science is one way of explaining our universe. Besides science, other ways of knowing include the humanities, belief systems, myths, and math. They are similar in that each has both a body of knowledge or information and its own processes for discovering truths, answering questions, and solving problems. Scientific thought arose as an alternative to, or more accurately a rejection of, ideas that were accepted because a ruler or religious authority declared their truth without empirical (meaning based on observation or experiment) evidence. This shift marked a revolution in our understanding of the natural world. It explained how the world works on the basis of observation and experimentation.

Thus, science differs from other ways of knowing in one essential way: It is the only form of discovery in which the process is the scientific method.

The Scientific Method

Students sometimes see the scientific method as an intimidating formal series of jargon-filled steps that scientists use to uncover truths or find answers. We have tried to abandon an arcane, philosophical approach and instead identify the scientific method's essence—the formation and rigorous objective testing of hypotheses.

■ Steps of the Scientific Method

Typically, the first step of the scientific method is observation. In this context, observation means more than merely seeing. It uses all of the human senses as well as the vast array of measurement techniques, some of which detect signals that human senses are incapable of receiving. Our knowledge of the Earth's core, mantle, and crust, for example, is based in part on the behavior of seismic waves, only some of which human senses can perceive, as discussed in Chapter 6, "Elements of Physical, Chemical, and Geological Oceanography."

Here's an important point: If an event or process cannot be observed, either directly or indirectly, it cannot be explained by science.

In the hands of a scientist, or even a curious lay person, observations lead to hypotheses, which can be educated guesses, carefully crafted explanations, or even questions. Consider the observation made first in the 1970s that Caribbean coral reefs were (and still are) suffering increased frequencies of disease and overgrowth by algae. It has long been known that the coral polyps, the living animals that form the calcareous foundation of the reef, are very sensitive to environmental factors such as temperature, salinity, light, nutrients, and sediment in the water.

Researchers noticed that the onset of the coral decline coincided with increasing aridity and desertification in northern Africa. Now, of course, this could be a coincidence. Their observation, however, led to a hypothesis. They proposed that dust from Africa resulting from drought and a loss of vegetative cover (desertification) was blown across the Atlantic Ocean by prevailing winds (coming from the northeast at those latitudes) and caused the coral problem, either through smothering the coral or by transportation of pathogens that harmed the coral. To formulate their hypothesis, they had to be familiar with the nature of global wind belts.

This story shows a key feature of the scientific method: Explanations can (and frequently must) be changed as new evidence becomes available.

A hypothesis is thus a proposed explanation of some phenomenon. The word phenomenon is from Greek *phainomenon*, meaning "to appear." It refers to any fact or experience that can be sensed and scientifically described.

After you have devised a hypothesis, the next step is to test it. Testing implies that your hypothesis can be falsified—that is, your hypothesis can be shown to be incorrect. Guesses, explanations, and questions that cannot be tested and falsified are—like things that cannot be observed—not science.

This "testability" requirement also means that science cannot make value judgments because qualities such as goodness or beauty cannot be subjected to scientific tests. Similarly, a scientist, acting solely as a scientist, cannot conclude that something is good or bad, beautiful or ugly, and so forth.

For the coral reef example, tests of the dust hypothesis were, in some cases, consistent with the hypothesis. Satellite photos showing dust clouds emanating from sub-Saharan Africa led to the calculation that several hundred million tonnes[1] of dust are transported over the Atlantic Ocean annually.

---

1. A *tonne*, as you will see throughout this book, is also known as a *metric tonne* and equals 1,000 kg, or 2,200 pounds. A *ton*, or *short ton*, equals 2,000 pounds.

Here are some important points about hypotheses. If the data[2] you collect support your hypothesis, then your hypothesis can be tentatively accepted. A hypothesis is never proven, as science does not deal in proofs, in spite of claims to the contrary in advertisements for consumer products. Moreover, acceptance of a hypothesis may be only temporary because scientists suspend judgment on making final determinations, except in very unusual situations. The original hypothesis may need to be revised or even abandoned completely as new information becomes available, new approaches to testing are undertaken, and/or as new technology becomes available. This means that scientists do not "jump to conclusions."

Science also demands that tests of hypotheses be repeatable. Thus, science advances very cautiously. If the data do not support a hypothesis, then it is rejected, no matter how good the hypothesis may have seemed or how much a scientist wanted to accept it.

■ Scientific Theories and Unifying Principles

Some hypotheses considered central to the understanding of a discipline—that is, some branch of science—have been subjected to an enormous amount of testing and having withstood this level of scrutiny have come to be regarded as scientific theories. In science, a theory is a broadly accepted explanation for an important phenomenon. This is a meaning that is very different from the nonscientific dictionary connotation of the word theory, which is simply conjecture, and implies considerable doubt. This is a very important distinction that should be carefully noted and remembered. A scientific theory could never be referred to as "only" a theory because extensive testing and retesting leave virtually no room for doubt. Plate tectonics and evolution represent theories in geology and biology, respectively.

These theories, plate tectonics and evolution, also represent unifying principles in their disciplines. A unifying principle is one that offers an overarching, or unifying, explanation for seemingly diverse phenomena and assembles them into a coherent whole.

■ Cause–Effect Versus Correlation

Let us revisit the African dust hypothesis that we used to explain the decline of Caribbean corals. The evidence obtained through scientific research seems to indicate three things: There were no coral health problems when dust concentration was low; the first appearance of African dust was associated with coral disease and death, and higher dust levels resulted in increased levels of disease and mortality.

This research shows a connection between dust levels and coral health. Such a comparison that shows a relationship between two variables—in the previously mentioned case, dust concentration and coral health—is known as a correlation. Correlations are important to the advancement of science and in many cases may be the only data available from which we can draw conclusions. Unfortunately, erroneous conclusions are also made from spurious correlations, especially in the media.

---

2. The word *data* is plural and thus takes the plural form of a verb. The singular form of *data*, that is, one piece of information, is known as a *datum*. In practice, the word *data* is sometimes used to denote the singular, but technically, this is incorrect.

A much stronger case for accepting a hypothesis can be made if there is a better link, preferably verified experimentally, between a cause and an effect.

Currently, a cause–effect relationship between coral health and African dust has not been firmly established, although the hypothesis remains a possible explanation for the observations. The possibility also remains that dust is one contributing factor in the decline of reefs, along with pollution, higher temperature, or some other factor or factors.

Science thus offers a way to understand our natural world that incorporates numerous "firewalls"—that is, phenomena must be observable, and hypotheses must be testable and falsifiable. Scientists also suspend judgment, and try to refrain from assessments of value.

In the following section, we examine how policy makers use scientific information.

Science and Public Policy: The Precautionary Principle and Scientific Uncertainty

At the start of the third millennium, there is an overwhelming consensus among atmospheric scientists that our growing human population is changing the composition of the planet's atmosphere. Even though scientists cannot yet be certain what the effect of these changes will be, the preponderance of evidence suggests that the impact will be, on the whole, negative and could be catastrophic for hundreds of millions of people crowded into the planet's coastal cities, as well as for entire ecosystems like coral reefs and mangroves. Although the present level of agreement among scientists has grown over the past decades to a consensus, there is still vigorous debate on the magnitude, timing, and nature of specific impacts.

Science, as you have learned, advances cautiously, in accordance with the principles of the scientific method. In the case of global warming and climate change, the scale of the phenomenon is so large and the subject so complex that achieving scientific certainty of the impacts is likely not possible. In this section, we introduce you to two approaches to applying science to policies such as climate change. They are the precautionary principle and the principle of scientific uncertainty.

Scientific certainty, as the preceding pages should have impressed on you, is very difficult to achieve. Even things we consider to be "laws" can be modified with new observations. Because the 100% level of certainty is thus not a practical threshold for accepting hypotheses, scientists typically use the 95% standard, which basically means that a hypothesis is accepted if 95% of the observations of a test are in line with the hypothesis, the other 5% varying because of random chance, experimental error, or some other factor. Although it may seem that lowering the level of confidence from 100% to 95% opens the door for hypotheses to be accepted when they are in fact wrong, setting a 100% level would result in virtually no scientific advancement. The 95% level of scientific certainty still allows science to advance cautiously, even if it is wrong as much as 5% of the time.

When it comes to artificial chemicals like the category called Persistent Organic Pollutants (POPs), scientists can rarely be 95% certain as to their impacts on individuals or ecosystems, as there are so many variables. In those cases, according to the "Rio Declaration" from the 1992 United Nations Conference on Environment and Development, "In order to protect the environment, the precautionary approach shall be widely applied. … Where there are threats of serious or irreversible damage, lack of full scientific certainty shall not be used as a reason for postponing cost-effective measures to prevent environmental degradation."

This standard is known as the precautionary principle.

Embedded in this principle is the notion that "there should be a reversal of the burden of proof, whereby the onus should now be on the operator or polluter to prove that an action *will not* cause harm, rather than on the environment to prove that harm (is occurring or) will occur."[3]

Another way to express this principle is "better safe than sorry." The products of science and technology are often brought to the marketplace without adequate investigation into any possible long-term effects on human health and the global environment. Some examples are the uses of POPs, lead, and mercury.

In most industrialized nations, the so-called burden of proof falls not on the producers of goods, but rather on those who allege that they have suffered harm. This is the basis of our tort system of civil law. As a result of the proliferation of new products, government agencies like the Food and Drug Administration, the Environmental Protection Agency, and the Federal Trade Commission, to name but a few, are sometimes unable to keep pace. For example, in deciding whether to approve pharmaceuticals for the market, the Food and Drug Administration purportedly adopts the precautionary principle in that it requires drug companies to demonstrate their products are safe and effective before they are sold. Even this approach, however, does not prevent many potentially harmful products to be marketed.

Although adult individuals have recourse to law if they believe they have been injured, fetuses, children, wildlife, and ecosystems have no such means of redress. Strict adherence to the precautionary principle could in the view of many facilitate democratic oversight.

Similarly, under the precautionary principle, a potentially serious threat such as global warming or the proliferation and buildup of organochlorines (a form of POP) in the ocean would trigger action to address the threat even if the science is not yet conclusive, but is supported by the preponderance of available evidence.

---

3. Glegg, G., and P. Johnston. 1994. The policy implications of effluent complexity. In *Proceedings of the Second International Conference on Environmental Pollution*, Vol. 1, p. 126. London: European Centre for Pollution Research.

CHAPTER 2

Principles of Critical Thinking

John Milton said this in *Paradise Lost:*
The mind is its own place
And in itself can make
A hell of heav'n
And a heav'n of hell.

It behooves us all to use the mind's power effectively, which involves critical thinking. Much of our thinking is spontaneous, is often emotional, and is rarely analytical and reflective. As such, it contains prejudice, bias, truth and error, inspiration, and distortions—in short, good and bad reasoning, all mixed together. Critical thinking essentially requires that we apply analysis, assessment, and the rules of logic to our thought processes.[1]

Scientists often equate critical thinking with the application of the scientific method, but we think critical thinking is a far broader and more complex process. Critical thinking involves developing skills that enable you to dissect an issue (analyze) and put it together (synthesize) so that interrelationships become apparent. It involves identifying assumptions, the basic ideas and concepts that guide our thoughts. Critical thinking also

---

1. Paul, R., and L. Elder. 2000. *Critical Thinking—Tools for Taking Charge of Your Learning and Your Life.* Upper Saddle River, NJ: Prentice Hall.

encourages an appreciation for our own and others' points of view, which is important when approaching complex environmental issues.

Too often, analyzing complex issues leads some to a belief that everyone is "entitled" to an opinion that should be respected. We do not necessarily concur; however, problem solving demands a willingness to listen for content to what others are saying. Talking is easy, but listening is not.

To develop critical thinking skills, all of us must learn to use a set of intellectual standards as an "inner voice" by which we constantly test and hone our reasoning, but the standards must be set in an appropriate framework in order for true critical assessment to take place.

The following describes in detail the intellectual standards we should apply when assessing the quality of our reasoning. This is the basis for critical thinking, which in turn is the approach that we try to apply throughout this book.[2]

Intellectual Standards: The Criteria of Solid Reasoning

Clarity. Clarity is the most important standard of critical thinking. If a statement is not clear, its accuracy or relevance cannot be assessed. For example, consider the following two questions:

1. What can we do about marine pollution?
2. What can citizens, regulators, and policy makers do to ensure that toxic emissions from industry, transportation, and power generation do not cause irreversible ecological damage to the marine environment, or harm human health?

Accuracy. How can we find out if a statement is true? A statement can be clear but not accurate.

Precision. A statement can be clear and accurate, but not precise. For example, we could say that "there are more personal watercraft in the United States in 2008 than ever before." That statement is clear, and it is accurate; however, how many more personal watercraft are there? 1? 1,000? 1,000,000? (There is a difference between the way many scientists use the word *precision* and the more general way it is used here.)

Relevance. How is the statement or evidence related to the issue we are discussing? A statement can be clear, accurate, and precise, but not relevant. Assume that we are given the responsibility to eliminate the harmful marine environmental impact of mercury emitted from coal-burning power plants, and we invite public comment on our proposals. Someone might say, "Electricity from coal-burning power plants

---

2. We obtained the basis for much of the information on critical thinking from International Conferences on Critical Thinking and Educational Reform sponsored by the Foundation and Center for Critical Thinking (www.criticalthinking.org). This section is adapted from *Environmental Issues: An Introduction to Sustainability* by Robert L. McConnell and Daniel C. Abel (2008, Pearson Prentice Hall, Upper Saddle River, NJ).

provides power for 100,000 jobs in this state alone." That statement may be clear, accurate, and precise, but it is not relevant to our specific responsibility of removing mercury.

Breadth. Are we considering all lines of evidence that could provide us with some insight in addressing an issue? Is there another way to look at this question? For example, in assessing the impact of African dust on coral reefs, we must also consider other causes, such as global warming, increased sedimentation due to deforestation of slopes on Caribbean Islands, and so forth.

Depth. Is a proposed solution realistic? How does it address the real complexities of an issue? This question is one of the most difficult to tackle because here is where reasoning, "instinct," and moral values may interact. The points of view of all who take part in the debate must be carefully considered. For example, politicians have offered the statement "just don't do it" as a solution to the problem of teenage drug use, including smoking. Is that a realistic solution to the problem, or is it a superficial approach? How would you defend your answer? Is your defense grounded in critical thinking?

Logic. Does one's conclusion clearly follow from the evidence? Why or why not? When a series of statements or thoughts are mutually reinforcing and when they exhibit the intellectual standards described previously here, we say they are logical. When the conclusion does not make logical sense, is internally contradictory, or not mutually reinforcing, it is not "logical." We give examples of logical fallacies later here.

■ Applying Intellectual Standards in a Critical Thinking Framework

The intellectual standards described previously are essential to critical evaluation of issues, but there are more factors to be considered. The following criteria constitute the framework in which these standards should be applied.

Point of view. What viewpoint does each contributor bring to the debate? Is it likely that someone who has a job on an ocean fishing fleet would have the same view on marine sanctuaries as someone who does not?

Identifying a point of view does not mean that the point of view should automatically be accepted or discounted. We should strive to identify our own point of view and the bases for this, we should seek other viewpoints and evaluate their relevance, and we should strive to be fair minded in our assessment. Few people are won over by having their opinions ridiculed. Furthermore, our points of view are often informed by our assumptions, which we address later here.

Evidence. All problem-solving is, or should be, based on evidence and factual information. Our conclusions or claims must be based on sufficient relevant evidence. The information must be laid out clearly. The evidence against our position must be evaluated, and we must be open to new evidence that challenges our conclusions.

Purpose. All thinking to solve problems has a purpose. It is important to have a clear understanding of that purpose and to ensure that all participants understand what that purpose is. Because it is easy to wander off the subject, it is advisable to check periodically to make sure that the discussion is still on target. For example, students working on a project occasionally stray into subjects that are irrelevant and unrelated, although they may be interesting or even seductive. It is vitally important, therefore, that the issue being addressed must be defined and understood as precisely as possible.

Assumptions. Here is an excerpt from a March 2008 report prepared by the US Energy Information Administration. "Growth in natural gas-fired generation is projected to be relatively flat [in 2008] due to the assumption that summer temperatures will fall back to near-normal levels." Assess the clarity and precision of this statement. This statement contains an assumption.

All reasoning and problem-solving depends on assumptions, which are statements accepted as true without proof. For example, students show up in class because they assume that their professor/teacher will be there. We should identify our assumptions, and always be ready to examine and evaluate them. They often need to be revised in the light of new evidence.

Now, before we analyze our own assumptions, let us summarize some characteristics of sound reasoning.

Critical thinking involves that we do the following:

- Continually exercise our thinking skills
- Can eliminate irrelevant topics and can explain why they are irrelevant
- Come to well-reasoned conclusions and solutions

Moreover, we should strive to understand concepts, key terms, and phrases essential to our discussion (such as greenhouse gas, ocean circulation, etc.).

Additionally, the effective reasoner continually assesses and reassesses the quality of his or her thinking in light of new evidence. Finally, one must be able to communicate effectively with others.

Summary

To summarize, the intellectual standards by which critical thinking is carried out are clarity, accuracy, precision, relevance, breadth, depth, and logic. These standards are applied in a framework delineated by points of view, assumptions, evidence or information, and purpose. We encourage you to return to this section whenever you need to refresh and polish your critical thinking skills.

CHAPTER 3

The Role of Government

Assumptions About Government's Role in Protecting the Environment

To assess the importance and power of assumptions in guiding your reasoning, take the following self-directed quiz.

First, identify your assumptions in defining government's responsibility to protect the environment. Second, determine your assumptions as to the proper level of government that may act. Third, determine your position on the "precautionary principle." Fourth, identify your assumptions concerning the extent to which individuals, governments, or institutions may impose costs on others with or without their knowledge or consent.

Thomas Jefferson and Government

Most politicians and many Americans probably consider themselves to have "Jeffersonian" principles. Read the following quotation taken from Thomas Jefferson's First Inaugural Address, delivered on March 4, 1801.

What more is necessary to make us a happy and prosperous people? Still one thing more, fellow citizens—a wise and frugal Government, <u>which shall restrain men from injuring one another</u>, shall leave them otherwise free to regulate their own pursuits of industry and improvements, and shall not take from the mouth of labor the bread it has earned.

QUESTION 3-1. In a clear sentence or two, explain what you think Jefferson meant by the italicized phrase. Now assess the breadth of your response by considering Questions 3-2, 3-3, and 3-4.

QUESTION 3-2. Do you think he was referring solely to thugs who physically brutalize their fellow citizens? Explain.

QUESTION 3-3. Could he logically also have been referring to citizens who sought to poison others? In other words, is restraining "poisoners" a legitimate role of government? Explain your answer.

QUESTION 3-4. Now, what if a citizen or organization dumps a toxin into water or air that all citizens depend on or if a citizen or organization fills in a wetland that performed valuable ecological functions on which local residents depend? May government under Jefferson's principle restrain that person or organization?

Your answer to these questions will define your assumptions as to the proper role of government. On what did you base your assumptions?

The Proper Level of Government That May Act

Next, evaluate your assumptions about the level of government, if any, which may properly intervene in environmental issues.

One of the major discoveries of the past 2 decades has been the extent to which much marine pollution is transboundary in nature. For example, one third of the air pollution affecting the Oregon coast comes from marine vessels outside the 5-km (3-mi) territorial limit controlled by the state, and more from power plants in Asia thousands of kilometers away. This situation is repeated over and over across the country and around the world.

QUESTION 3-5. In the light of the transboundary nature of pollution, is it appropriate that local government by itself (e.g., the coastal city of Newport, Oregon) bear the responsibility for protecting its own environment? May the states and federal government have a legitimate role? Should international agencies be involved? Explain and justify your answer.

Your answer will help evaluate your assumptions about the extent to which state, federal, or global agencies have responsibilities to intervene to protect local environments.

The Question of Externalities

Economists define externalities as any cost of production not included in the price of the good. An example would be environmental pollution or health costs resulting from burning diesel fuel, not included in the price of the fuel. Another example is the cleanup costs paid by governments resulting from animal waste pollution of water bodies from large-scale meat-processing operations. In this example, the price of chicken or pork at your local supermarket is lower than it would be if all environmental cleanup costs were included in the price of the meat.

Assumptions About Corporations

Here is another choice quotation from Thomas Jefferson on the impact of those new organizations called corporations. Read Jefferson's words, and then respond to the following question:

> *I hope we shall take warning from, and example of, England, and crush in its birth the aristocracy of our moneyed corporations, which dare already to challenge our Government to trial, and bid defiance to the laws of our country.*

QUESTION 3-6. Do you share or reject Jefferson's opinions concerning corporations? Justify your conclusion. Prepare a list of positive and negative contributions corporations make to our environment and economy. Do you conclude corporations have "too much power" in contemporary life? Why or why not?

If you are interested in corporate power, research the 1886 US Supreme Court ruling: *Santa Clara County v. Southern Pacific*. The Court ruled that Southern Pacific was a "natural person" entitled to the protections of the US Constitutions Bill of Rights and 14th Amendment. After researching the case, answer Question 3-7.

QUESTION 3-7. Do you believe the Court acted correctly in deciding that a corporation was a person? Is the 1886 ruling relevant to the 21st century? Why or why not?

After having thoughtfully responded to the previous scenarios, you should now have a better awareness of the assumptions that you bring to the analysis of environmental issues that you are about to undertake in *Environmental Oceanography*.

Knowledge, Opinions, and Logical Fallacies

Knowledge and Opinion

The philosopher Mortimer Adler pointed out[1] that there is no contradiction in the phrase "true opinion," nor is it redundant to speak of "false opinion." Thinking about the subject matter of this book requires that we separate truth and knowledge from opinion and false opinion from true opinion.

There are few things that are both incorrigible and immutable: One example is the statement that the sum of a finite whole is greater than any of its parts. Sometimes definitions are self-evident truths: For example, a triangle has three sides.

Few scientific concepts meet these rigid standards. Does that mean that everything else is opinion? That depends on how we define opinions. Some opinions deserve the status of knowledge, even if they are not immutable and incorrigible. Examples are theories that have an overwhelming body of supporting evidence. We can say with confidence that such ideas are true at a particular time. Should new evidence come to light, we may have to evaluate our theory.

---

1. Adler, M. J. 1985. *Ten Philosophical Mistakes.* New York: Macmillan.

Contrast these kinds of opinions with personal prejudices, which we assert often without any evidence or force of reason to support them. We may, for example, feel that "the government has no business telling people what to do with their land." This is what philosophers would call a "mere" opinion, as opposed to a "true" opinion, which we illustrated earlier. There is nothing "wrong" with having mere opinions, but we all should recognize them for what they are.

Common Logical Fallacies

Much of your information concerning marine issues will come from the media—television, magazines, radio talk shows, and newspapers. These sources often exhibit evidences of poor reasoning, such as logical fallacies (as well as mere opinions). Learn to identify them to ensure that you are getting the best information possible. Here, in no particular order, are some of the more common examples of logical fallacies:

- The fallacy of composition: assuming that what is good for an individual is good for a group. An example is standing at sports events—this is an advantage to one person but not when everybody does it.

- The fallacy of starting with the answer: including your conclusion in your premises or assumptions. For example, "America runs on high levels of energy consumption. We can't have the American lifestyle without energy. Thus, America can't afford to cut energy use." Here, the arguer is simply defining his way out of the problem. By his reasoning, we would have to continue to increase energy consumption forever, an obvious impossibility.

- The fallacy of hasty generalization: "Senegal and Mali have very low levels of energy consumption. They are very poor countries. Low levels of energy consumption lead to poverty." How is poverty measured? Are there any "wealthy" countries that have relatively low levels of energy consumption? Are there any relatively poor countries that have high levels of energy consumption?

- The fallacy of false choice: stating an issue as a simplistic "either–or" choice when there are other more logical possibilities. "Those who don't support fossil fuel use want to go back to living in caves."

- Fallacy of an appeal to deference: accepting an argument because someone famous supports it.

- Fallacy of *ad hominem* argument (literally, "at the person"): attacking a person (or his or her motives) who advocates a position without discussing the merits of the position.

- The fallacy of repetition—the basis of most advertising: repeating a statement without offering any evidence. "Population growth is good. People contribute to society. We need population growth to survive."

- The fallacy of appealing to tradition: "Coal built this country. Eliminating coal use would threaten our society."

- The fallacy of appealing to pity: "The commercial fishing industry supports millions of American families. Regardless of its impact, we have to support them."

- The fallacy of an appeal to popularity: "Seventy five percent of Americans support this position." Perhaps the poll asked the wrong question. Perhaps the respondents did not have enough information to properly respond, or perhaps they were uninformed, and so forth.

- The fallacy of confusing coincidence with causality: "After passage of the Endangered Species Act, jobs in sawmills fell 70%; therefore, the Endangered Species Act was bad for the economy." Were there other possible explanations for the drop in jobs?

- The fallacy of the rigid rule: "Hard-working people are good for the economy. Immigrants are hard-working people; therefore, the more immigrants we have, the better for our economy." "Large numbers of immigrants commit crimes. Crimes are bad for the economy; therefore immigration should be curtailed."

- The fallacy of irrelevant conclusion: using unrelated evidence or premises to support a conclusion. "Development raises the value of land and provides jobs. Developed land pays more taxes than undeveloped land; therefore, any available land should be developed."

PART 2

Principles of Oceanography

CHAPTER 5

Towards Sustainable Oceans

In 1992, 1700 scientists, including most living Nobel laureates in the sciences, issued the World Scientists' Warning to Humanity: Human beings and the natural world are on a collision course. Human activities inflict severe and sometimes irreversible damage on the environment and on critical resources. If unchecked, many of our current practices put at serious risk the future that we wish for human society and life on Earth and may so alter the planet that it will be unable to sustain life in the diversity that we now know. Fundamental changes are urgent, they conclude, if we are to avoid the damage our present course will bring about.

The solution to these environmental problems is sustainability.

What Is Sustainability?

Sustainability, or sustainable development, was defined in 1987 by the World Commission on Environment and Development[1] (also known as the Brundtland Commission, named after its chair, Gro Harlem Brundtland) as development that meets the needs of the present without compromising the ability of future generations to meet their own needs. Sustainability transcends and in many ways supersedes environmentalism. It involves a transformation from a wasteful linear model of resource use in which natural resources (be they living or nonliving) are extracted, used, then thrown "away" to a cyclical model built around reduction, reuse, and recycling.

Quantifying Sustainability

In 2005, the Environmental Performance Measurement Project at Yale University issued an Environmental Sustainability Index (ESI), ranking nations on the extent to which their societies approached sustainability. "No country is on a sustainable trajectory— and the ESI demonstrates this," said Gus Speth, Dean of the Yale School of Forestry and Environmental Studies. In most cases, insufficient data exist to determine each nation's ESI accurately. Their rankings are, therefore, only crude comparative measurements of societal sustainability. For national rankings and details, go to www.yale.edu/esi/.

The Sustainable Communities Network sets general targets for a Sustainable Planet Earth. They are as follows:

- Creating community
- Growing a sustainable economy
- Protecting natural resources
- Governing sustainably
- Living sustainably

---

1. World Commission on Environment and Development (Gro Harlem Brundtland, Chair). 1987. *Our Common Future.* Oxford: Oxford University Press.

• •

Achieving a Sustainable Marine Environment

In our view, achieving a sustainable marine environment requires addressing a number of seemingly unrelated issues. These include fisheries, forests, energy, biodiversity, climate change, manufacturing and industry, and justice and equity.

■ Fisheries

Eventually, the impact of our growing demand for fish protein must be addressed. Contemporary industrial-style fishing methods impose significant environmental costs. For example, deep-ocean trawling has been compared with clearcutting a forest, only on a much larger scale. Moreover, per capita yields from global fisheries are declining, and expansion of aquaculture often occurs at the expense of coastal ecosystems and wild fish stocks (see Issues 12, 13 and 17–21).

Many fish species are in precipitous decline. Thriving and diverse aquatic wildlife are necessary for healthy marine and freshwater ecosystems (see Chapter 6). It is therefore critical that those dependent on fisheries and aquatic ecosystems use these resources responsibly. This can mean creating marine protected zones that are large enough to maintain species diversity, away from human interference; however, protected zones alone may not be enough. Large fish and marine mammals characteristically have dangerously high levels of toxic artificial chemicals in their tissues, from pollution sources perhaps thousands of kilometers away.

Protecting aquatic wildlife could be aided through sustainable aquaculture. For example, growing plant-eating (herbivorous) fish such as carp and tilapia puts less strain on resources compared with growing carnivorous species like salmon, which usually must be fed feed made with wild-caught fish.

Shrimp farming puts great stress on coastal ecosystems, as mangrove communities are often destroyed to make room for shrimp ponds, especially in developing countries such as Honduras and Vietnam, which in turn export the farmed shrimp to the United States. Populations of many oceanic species are at critical levels because of industrial-style fishing practices as well as massive national subsidies for fishing fleets. Ironically, the destruction of ocean fisheries coincides with an increased demand for fish resulting from its recognition as a health food.

■ Protecting Terrestrial Ecosystems

Although trees have economic value as a raw material, the environmental services provided by forests far transcend the economic value of trees in many cases. Mature trees maintain desirable microclimates and retard sediment loss that can poison near-shore marine ecosystems such as coral reefs. Forests store carbon, mediating climate change and thereby reducing excess carbon dioxide dissolved in seawater, which can lead to acidification of the oceans.

▪ Energy

Sustainable societies cannot be built on nonrenewable energy resources. Humans use almost unimaginable amounts of energy and generate vast amounts of pollution from it. Fossil fuel burning emits pollutants like SOx and NOx (oxides of sulfur and nitrogen), particulates, and heavy metals such as mercury. These have now been distributed throughout the oceans from airborne fossil fuel emissions.

Moreover, the transport of petroleum by tanker inevitably results in oil spills that degrade and may destroy sensitive marine habitats. Energy conservation and the use of renewable fuels provide cost-effective and sustainable alternatives that generate little if any air and water pollution. Subsidizing the production of coal and oil-based nonrenewable energy makes little sense in a world threatened with rapid climate change and accelerating species loss. Here, too, change is coming. Wind energy is the fastest growing energy source in Europe and North America, supported by government subsidies, which partly offset subsidies for fossil fuels. Large-scale development of offshore "wind farms" is planned for Britain and elsewhere in Europe. Such developments, while offering virtually pollution-free energy, must be carefully constructed and monitored so as to avoid harming the local marine environment. Likewise, offshore oil and gas development, a fixture in the oceans since the 1950s, must be restricted and, where conducted, more carefully undertaken in marine environments already stressed by overfishing, climate change, sediment pollution, and so forth.

▪ Biodiversity

Our very survival could ultimately rely on maintaining the integrity of marine and terrestrial ecosystems that we barely understand. An ecosystem is a geographic area that includes all living organisms with their physical surroundings and the natural cycles that sustain them (such as the hydrologic cycle). All these elements are interconnected. Altering any one component affects the others in a particular ecosystem. Biodiversity, the mix of organisms within an ecosystem, is particularly critical for sustainability because of the specialized and often little-understood roles each species plays in maintaining the dynamic state of ecological balance. Moreover, little is known about key ecosystems like soils and the deep ocean.

Esthetics and ethics must also play a part, as humans can probably survive on an Earth with drastically reduced species diversity. The question then becomes this: Do we wish to make a decision to eradicate species and ecosystems without the input of our descendants? That our ancestors did so in ignorance is no excuse for our perpetuating such behavior.

▪ Climate Change

We consider climate change more fully in Issues 4–7. Although climate change is a well-documented fact of planetary history—the Earth has gone through several megacycles of "greenhouse" and "icehouse" conditions—the speed with which human-induced climate change is occurring is virtually unprecedented. Too-rapid change overwhelms the ability of natural ecosystems to adapt. For example, rising sea levels may flood and destroy coastal salt marshes if sea level rise is too fast for these communities to migrate inland, and rising shallow-water ocean temperatures can lower oxygen content and threaten communities like coral reefs already operating throughout the oceans at temperatures near their thermal maxima. The impacts of climate change are imperfectly understood, but on the whole will test the capability of a human species that "subdued" a seemingly limitless Earth. Moving to sustainable societies may be essential to address the impacts of climate change.

■ Manufacturing and Industry

The Industrial Revolution generated wealth beyond humanity's dreams but also generated waste in unprecedented quantities, some of it artificial chemicals never before seen on Earth. Processing this waste is rapidly exceeding the capacity of natural systems, which did not evolve in the presence of many of these chemicals.

In nature, waste eventually becomes something else's food. In human societies, waste is everywhere, and it indicates inefficiency. Pollution is one form of waste. Approaching "wasteless" production must become the norm in human activity, as it is in the marine realm, and progress is being made: The European Union has set a goal of ending landfill disposal by 2025. Many businesses have found that waste reduction and even elimination can enhance profitability, and progress has been made here as well—humans have agreed to phase out or eliminate the most harmful kinds of Persistent Organic Pollutants (POPs; see Issue 8); however, the growth of human populations and the universal association between increasing wealth and increasing waste pose critical problems for a world, five sixths of whose population is trying to develop along Western-style free-market lines.

■ Justice and Equity

The pursuit of justice and equal opportunity are key ingredients in building a sustainable society. Examples of injustice are a lack of adequate housing, a lack of access to education, poor sanitation, an inadequate supply of pure water, exposure to environmental toxins, and environmental degradation related to industrial pollution. Rich societies ignore these issues at their peril. Furthermore, injustice often drives rapacious and unsustainable use of natural resources such as coral reef fisheries and mangrove wetlands.

. . . .

Sustainable Consumption: An Oxymoron?

We are in the twilight of the era of the "myth of unlimited resources." Sustainability involves adopting responsible patterns of buying, consumption, and reproduction, thereby consuming minimal energy and fewer resources. Responsible consumption is based on education, not coercion, in a democratic society. Unfortunately, industrial and postindustrial societies are philosophically based on the myth of ever-increasing consumption, in turn driving ever-increasing production—the "growth" concept. Detoxifying society from this "unlimited consumption" myth may be one of our greatest challenges.

Sustainable Population

Human numbers must eventually become stabilized, as it is physically impossible for population growth to continue forever. The only questions are at what level will growth end and whether growth will end as a result of human actions or by natural processes like famine, disease, and war. Ageing societies are typical of developed nations. Populations dominated by the young are typical of developing ones. Large numbers of young people provide great promise for societies, but also impose great costs. The readers of this book are mainly young, and it is they who will address, or not address, these challenges. The next century should prove to be one of the most interesting and potentially rewarding centuries in the entire span of human history.

Development

Here we use development to refer to a complex set of changes that convert the economy of a society based on subsistence agriculture to one in which most of the employed inhabitants work in manufacturing or services.

Here are two hypotheses, much simplified, that purport to explain such relationships as may exist between development and the environment.

1. *Development harms the environment.* Many environmentalists assert that development leads to some or all of the following: destructive land-use practices, mining of marine resources, injurious levels of air emissions and fossil-fuel use, and water pollution. Moreover, high levels of population growth in many developing countries exacerbate environmental degradation, leading to misery, child prostitution, civil wars, and the like and encouraging large-scale emigration.

2. *Development eventually improves the environment.* Many economists and some environmentalists, while acknowledging harmful levels of environmental pollution in countries in early stages of development, cite considerable empirical evidence that (1) population growth rates decline as development proceeds and (2) rates of some forms of environmental pollution decline as per capita income increases, a supposed corollary of development, as we noted previously. Newer forms of technology tend to be less polluting than older forms, but also tend to require high capital expenditures. Countries with higher per capita incomes tend to have cleaner environments along with increased consumption of goods and services. As nations become richer and middle classes expand, demands for scarce goods such as wild fish stocks expand, and tougher environmental standards may also; however, in many cases, passage of comprehensive legislation to preserve the coastal environment in a rich country simply leads to mining of marine resources

for export from poorer countries. Most recently, the decline in fish stocks from the Atlantic off Northwest Africa has been due to increased exports to Europe and a concomitant decline in traditional fisheries (see Issue 18 for details).

Global Trade and the Oceans

Here are some examples of the adverse effects of global trade on the marine environment:

- Moving oil by tanker leads to oil spills, and the "cleanup" costs are not included in the price of oil.
- Cruise ships are a major source of untreated human and other waste (see Issue 9), often dumped at sea.
- The introduction of invasive species into new environments (see Issues 22 and 23) typically occurs by transfer of nonnative species in the ballast water of cargo ships.

The fuel of marine vessels is typically #6 fuel oil, which contains the highest levels of polluting sulfur of any form of petroleum-based fuel. Operation of marine vessels in ports during loading and unloading of cargo is often a major source of local air pollution. A good example is the Long Beach-Los Angeles, California harbor complex.

Such activities represent vast subsidies to world trade. In other words, were the traders required to pay all environmental costs associated with their activities rather than dump those costs onto the environment of the receiving country, the volumes and patterns of world trade would no doubt be considerably different.

Addressing Environmental Impact at the International Level

In the case of pollution that crosses national boundaries, addressing the problem usually means international agreements, and we give three examples here.

■ The World Trade Organization

In 1999, Seattle, Washington was the site of sometimes-violent protests against perceived policies of the World Trade Organization (WTO). We cite one example to illustrate the controversy: the issue of sea turtles. Five Asian nations challenged a US law designed to protect sea turtles from certain harmful fishing practices. The law banned the importation of shrimp from countries that did not require the use of turtle-excluder devices by their fishing industry. The WTO ruled in favor of the Asian countries, not because it disapproved of United States attempts to protect sea turtles but because the panel found the United States had discriminated among members of the WTO by granting preferential treatment to Latin American and Caribbean nations. The WTO decision infuriated environmentalists, even though it had arguably nothing to do with the desirability of saving sea turtles. Indeed, the founding charter of the WTO formally addresses the relationship of trade to the environment. Signatories to the WTO should "allow for the optimal use of the world's resources in accordance with the objective of sustainable development, seeking to both protect and preserve the environment." WTO rules also allow countries to impose trade regulations "necessary to protect human, animal, or plant life or health" or "relating to the conservation of natural resources"; however, measures taken to protect the environment must not discriminate. A country may not be lenient with its domestic producers and at the same time be strict with foreign producers nor can member nations discriminate among different trading partners. Sovereign nations choose to become members of the WTO and to act by its rules. So far, more than 140 countries have joined, and others have applied for membership.

■ The Montreal Protocol

Severe depletion of stratospheric ozone has been measured for years, especially in the Southern Ocean around Antarctica. The Montreal Protocol on Substances that Deplete the Ozone Layer was adopted in 1987 to eliminate the production and consumption of ozone-depleting chemicals. Four agencies were tasked with implementing the Protocol: the World Bank, the U.N. Environment Programme, the U.N. Development Programme, and the U.N. Industrial Development Programme. The Montreal Protocol stipulates that the production and consumption of compounds that deplete ozone in the stratosphere—chlorofluorocarbons, halons, carbon tetrachloride, among others—were to be phased out between 2000 and 2005. Developing countries were given time to cap and then eliminate ozone-depleting chemicals.

■ The Law of the Sea Treaty

Passed in 1982, the U.N. Convention on the Law of the Sea became effective in 1994. More than 150 nations and the European Community have ratified it. The United States has signed the treaty, but as of 2008, the Senate had not ratified it.

The treaty, among other things, defined certain territorial limits to which nations may exert influence. It further established EEZs, or Exclusive Economic Zones, within which nations had exclusive rights to exploit natural resources. It established a general framework for addressing environmental degradation in the oceans and set rules for exploitation of seabed resources outside of any nation's EEZ. It provided for access to the sea for landlocked nations.

■ The Next Steps

To eliminate the adverse environmental impacts of global trade on the world's oceans, many environmental scientists recommend additional international or multilateral agreements. They include proposals to

- Eliminate invasive species from the ballast water of vessels using international waterways
- Regulate international transboundary air and water pollution

- Require marine vessels to use low-sulfur fuels
- Require "double-hulled" tankers for the shipment of petroleum—already required for shipment of oil in United States territorial waters
- Provide that industries that exploit marine resources must adhere to a similar set of environmental regulations globally

■ A Final Word

International maritime commerce has enormous potential to foster the objectives of development but can also be the source of serious environmental degradation, absent multinational, and international agreements to "level the playing field." The Montreal Protocol and U.N. Convention on the Law of the Sea are examples of the types of international agreements that could serve as templates for new initiatives.

CHAPTER 6

Elements of Physical, Chemical, and Geological Oceanography

CHAPTER OUTLINE

continued on next page

Thinking Like a Scientist

Scientists use hypotheses to explain some natural phenomenon and then gather evidence to test their hypothesis. A hypothesis is an educated guess that is supported by some evidence and that can be tested by further observation and/or experimentation. If the hypothesis is upheld, scientists may elevate it to a theory.

In contrast to its meaning among nonscientists ("a theory is just a guess"), a theory is a powerful concept, supported by a vast array of observations and tests. Theories can be modified, and often are, by additional data.

Scientists often use models to explain the workings of some natural system. We illustrate by discussing the nature of the Earth's interior, which you need to know to understand how the oceans were formed and how they change over time.

Modeling the Earth's Interior

A model of the Earth's interior shows its general structure such as (1) interior layers drawn to scale, (2) where interior boundaries are, (3) what the states of matter are (whether solid, liquid, or gas), and (4) the composition of the interior, at whatever scale and in as much detail as is supported by evidence. We may not know the exact composition of a particular zone, or the layers may vary in composition, and so forth; thus, our model might try to show an "average" composition, or our model might try to show how compositions vary with depth.

To begin to make our model, we use all available evidence that will allow us to put constraints on what our model looks like. For example, we know from gravitational interactions between the Earth and other bodies in the solar system that the Earth's overall specific gravity, a measure of density or "heaviness," is 5.52, and thus, the specific gravity of Earth's interior material must average 5.52.

■ Summary of Evidence Used to Model the Earth's Interior

We use evidence from the following sources to make a model of the Earth's interior.

- We have excellent knowledge of the composition and distribution of surface rocks. (There are three categories of rocks: igneous, which form from a molten mass called magma; sedimentary, which form by erosion of pre-existing rocks or by organisms, like corals; and metamorphic, a "hybrid" rock formed when a pre-existing rock is subjected to heat and pressure, causing the original minerals to align themselves, recrystallize, or both.) Besides surface exposures of rocks (outcrops), we have hundreds of thousands of samples of the shallow crust (see p. 32–33) from bore holes, and thus, we can estimate the average composition of the Earth's continents and ocean basins.

- We use evidence from meteorites—the "raw material" of the original Earth. Meteorites are commonly either an iron-nickel alloy (approximately 6%) or are composed of silicate (containing the elements silicon and oxygen) minerals (approximately 93%).

- We have thousands of drill holes and dredge samples of oceanic sediment and underlying bedrock, and thus, we know the near-surface composition very well. Moreover, we have very reliable seismic data that show us where phase boundaries exist in the oceanic crust. We can confidently infer the composition of these layers. Figure 6-1 shows the structure of the upper 500 km of the Earth, including the oceanic and continental crust.

- As stated earlier, knowing that the Earth's average density is 5.52 puts strong constraints on what substances can make up the interior.

- The physics of the Earth's rotation dictates that the densest (heaviest) material must be at the Earth's center, and densities must decrease outward (and, of course, we know what the average density of the outer layer is from direct samples).

- Seismic waves generated by earthquakes can travel through the entire Earth. Like light off a mirror, seismic waves can bounce off surfaces inside the Earth where densities abruptly change. Thus we know where the major phase boundaries are between sub-surface materials, and we know something about their properties, as rock properties determine seismic wave velocity. For example, other things being equal, the higher the density of rock, the faster seismic waves travel through it. Figure 6-2 shows how seismic waves are reflected and bent (refracted) by the various layers inside the Earth. Internal

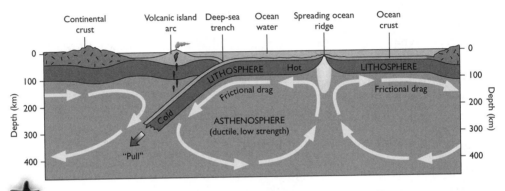

FIGURE 6-1 Structure of the upper 500 km of the Earth. The arrows represent convection "cells." The movement of hot, plastic asthenosphere material tugs at the colder, brittle lithosphere (see p. 33), ultimately fracturing it, at the "Spreading ocean ridge." Arrows show movement of material.

MIT news

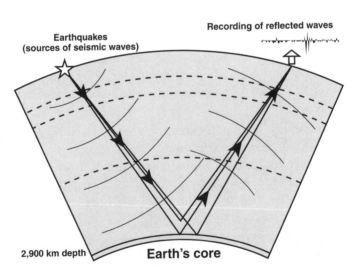

FIGURE 6-2 Seismic waves bounced off the Earth's core. The star represents the location of an earthquake. The pattern above the arrow represents the recording of the arrival of the waves at a seismograph station.

zones where waves abruptly change velocity are known as discontinuities. They are important internal boundaries and indicate a change of state, composition, or both.

- We know a great deal about the Earth's powerful magnetic field (Figure 6-3), and its very existence puts limits on the possible composition and state of the planet's interior.

Using such information, geologists have confidently constructed a general model of the interior, mindful, however, that new data may cause us to alter our model. This model can then be used to explain the origin and development of ocean basins.

Normal Polarity

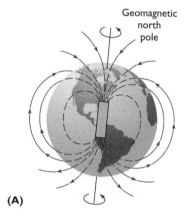

(A)

Reverse Polarity

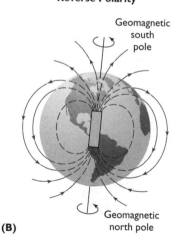

(B)

FIGURE 6-3 The Earth's magnetic field. The orientation of the lines of force indicates the inclination of a magnet at the Earth's surface. At the "geomagnetic equator" a magnet would be parallel to the surface. Near the north magnetic pole a magnet would be steeply inclined at an angle, if free to move. (A) Normal polarity, showing present orientation of lines of force. (B) Reversed polarity, showing that the magnetic field has "flipped" polarity in the geologic past. [*Source:* Adapted from A. Cox, *Science* 163 (1969): 237-45.]

Structure of the Earth: Lithosphere, Asthenosphere, Crust, Mantle, Core

One of the most fundamental discoveries about the Earth was that the interior is composed of layers with recognizable boundaries. Geologists at first divided the Earth into three layers: the crust (outermost material), an underlying mantle, and a core at

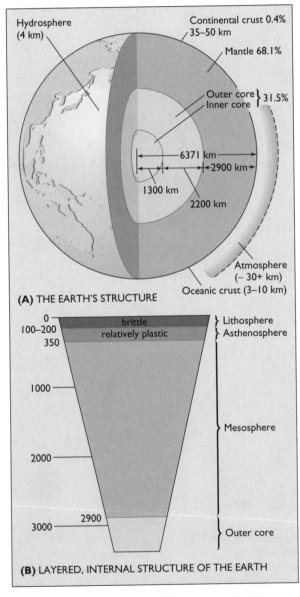

(A) THE EARTH'S STRUCTURE

(B) LAYERED, INTERNAL STRUCTURE OF THE EARTH

FIGURE 6-4 The Earth's internal structure. Percentages indicate the relative masses of the core, mantle and continental crust. Where is most of the Earth's mass concentrated?

the center, based on hypothesized differences in rock composition. Figure 6-4 shows the major zones within the Earth. Material may vary in composition, properties such as strength (which measures how it responds to deforming force), or state (whether solid, liquid, or gas). Strength and state in turn depend on the temperature and pressure to which the material is subjected.

Later, study of seismic waves and the Earth's magnetic field, coupled with laboratory research, allowed detail to be added to this model. Based on detailed measurements of the behavior of seismic waves, the core was hypothesized to consist of an inner solid portion and an outer liquid portion. The lower mantle had to be denser than the upper mantle. The upper mantle contains a zone near its top where seismic wave velocities were lower than those above and below—this zone became known as the low-velocity zone, or asthenosphere (from Greek *asthenos*, meaning weakness or loss of strength). The asthenosphere is a weak, plastic (meaning deformable), partially molten zone, distinguished from the lithosphere (Greek, *lithos* = rock), a strong, rigid zone, above it (Figures 6-1 and 6-4). The lithosphere sits atop the asthenosphere and includes the crust (both continental and oceanic) and the rigid uppermost portion of the mantle.

The upper mantle's composition is best described as peridotite. Peridotite takes its name from the gem-quality form of the mineral olivine—peridot, which is a magnesium-rich silicate mineral. Silicon is the second most abundant element in the Earth after oxygen. It should be no surprise then that minerals made mainly of silicon and oxygen should be so common.

Summary of the Earth's Interior

Our model at present is like this: The Earth is a layered body that has discrete boundaries separating the interior layers. These layers, based on compositional differences from the surface inward, are, crust, mantle, and core. The uppermost few hundred kilometers of the Earth are divided into a lithosphere (containing the crust and uppermost mantle) and an asthenosphere, or low-velocity zone, a partially molten zone entirely within the mantle. The crust is of two types: oceanic, consisting mainly of basaltic rocks below a variable layer of sediment, and continental, composed of all three rock types (igneous, sedimentary, and metamorphic), but on average having a composition near that of granite. The outer core is molten and the inner core solid. The composition of the core is mainly iron, with some nickel and minor elements such as sulfur, silicon, and/or potassium. The upper mantle is of peridotitic composition, and the lower mantle is probably made of densely packed oxide minerals with a similar composition to the upper mantle, but made of denser crystal structures.

Radioactivity and the Dynamic Earth

Radioactivity refers to the spontaneous "decay" of unstable atomic nuclei. Heat, a major by-product of radioactive decay, powers the global process that we call plate tectonics. The transfer of heat by convection (currents in the fluid caused by temperature differences) can produce convection cells, like those in a pot of soup that constantly bring material from the bottom of the pot to the top. Sometimes candle-like plumes of red-hot material form by convection. Convection cells and plumes powered by radioactive decay are the means by which new oceanic bedrock is made, and new oceans may be formed in the process.

The Dynamic Earth Model-Global Tectonics

■ Plates and Plate Boundaries

Plates are segments of the lithosphere (Figure 6-5). Plates are generally very thin (tens of kilometers is typical) relative to their other dimensions; therefore, to scale, they are comparable to the fractured shell of an egg.

There are three ways that plates can interact on the surface of a sphere like the Earth. They can move away from each other (diverge). They can move towards each other (converge), and they can slide past one another (called strike slip, or shear).

The three types of plate boundaries are therefore divergent, convergent, and shear (Table 6-1).

Divergent Boundaries

Ascending convection cells, consisting of a molten mush with crystals, spread out at the base of the lithosphere, heating it, tugging at its base, and weakening it. With enough time, the spreading convection cell may fracture the now hot, thinned, and weakened overlying

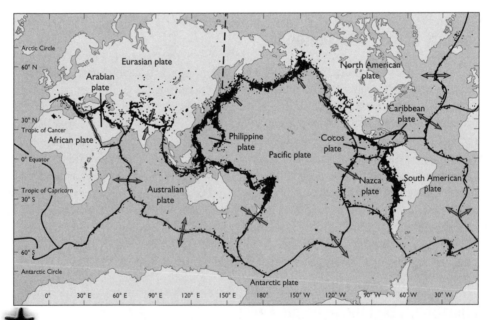

FIGURE 6-5 Plates of the Lithosphere. Clusters of black dots indicate earthquake activity. Arrows show relative plate motion. Arrows pointing away indicate tensional force at that boundary. Arrows pointing towards each other indicate compression. [*Source:* Adapted from Stowe, K.S. *Ocean Science*. John Wiley & Sons, Ltd., 1983.]

TABLE 6-1	Types and attributes of plate boundaries			
Type	Relative Plate Motion	Topography	Earthquakes	Volcanism	Examples
Midocean ridge	Divergent	Ocean ridge	Shallow, weak to moderate intensity	Volcanoes and lava flows	Mid-Atlantic Ridge
Subduction zone	Convergent				
	Ocean–ocean collision	Deep-sea trench and volcanic arc	Shallow to deep, weak to strong intensity	Volcanic arcs	Aleutian Islands
	Ocean–continent collision	Deep-sea trench and volcanoes	Shallow to deep, weak to strong intensity	Volcanic arcs	Andes Mountains
	Continent–continent collision	Mountain belt	Shallow to deep, weak to strong intensity	None	Himalayan Mountains
Transform fault	Strike slip	Offset ridge crest	Shallow, weak to strong intensity	None	San Andreas Fault

lithosphere. Pressure differences force magma upward, sometimes reaching the sea floor, where the lava freezes (congeals) and forms new oceanic bedrock. In this manner, new oceanic crust is formed. Extensive fracture systems thousands of kilometers long eventually develop and result in divergent plate boundaries. This is how new oceans are formed.

Convergent Boundaries

At convergent boundaries, old oceanic lithosphere that formed at divergent boundaries is recycled back into the mantle. Figure 6-6 shows an ocean–continent convergent boundary (discussed later here).

Convergent boundaries may be of three types. If two masses of continental crust are brought together at a subduction zone, a continent–continent convergent boundary is formed. This is how continents have been assembled over the past 2.5+ billion years. The Himalayas formed when India collided with Asia (Figure 6-6), nearly doubling the thickness of the continental crust there.

If oceanic lithosphere is subducted beneath continental lithosphere, an ocean–continent convergent boundary is formed (Figure 6-6), and if two oceanic lithosphere plates converge, an ocean–ocean convergent boundary is formed.

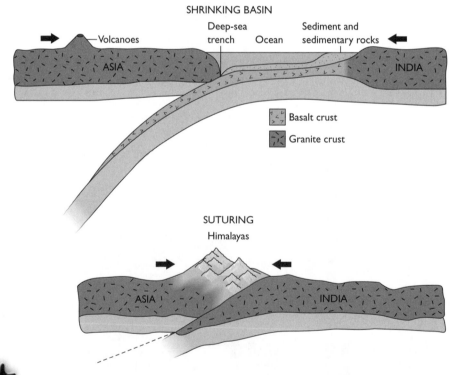

FIGURE 6-6 Ocean-Continent Convergent Boundary. The Himalayas are built of marine sedimentary rocks, which are less dense than the basaltic crust on which they rest. They cannot therefore be easily "subducted" beneath Asia and so pile up between the colliding masses of continental crust.

Convergent boundaries have distinctive "signatures," including trenches, island arcs, and patterns of earthquakes. Trenches form by the incessant downward tugging on the oceanic lithosphere. Earthquakes are distributed along subduction zones from near the surface (shallow-focus), through intermediate depths (intermediate-focus) to depths where the subducting material gets too weak to store stress (about 700 km). Deepest earthquakes are called deep-focus earthquakes. Island arcs, arcs of volcanic islands formed by subduction at ocean–ocean boundaries, are one of the planet's most distinctive features. Japan, the Aleutians, and Indonesia are examples. Figure 6-7 shows an island arc with its associated trench, in this case Indonesia. This was the site of the devastating 2004 Sumatra-Andaman earthquake and tsunami, which killed over 225,000 people. Island arcs are also the site of some of the world's highest concentrations of human populations and some of the most severe geohazard zones.

Shear Boundaries

When plates slide laterally past one another, transform or shear boundaries result. Transform boundaries on continents are recognized by strike-slip faults like the San Andreas Fault in California, the Alpine Fault in New Zealand, and the Dead Sea Fault Zone in the Middle East (Figure 6-8).

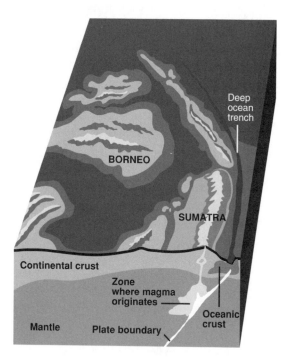

FIGURE 6-7 The Indonesian Island Arc. Here the plate boundary is steep, and the friction between the sliding plates causes the basaltic rock to melt, resulting in widespread Indonesian volcanic activity, concentrated in the southern part of Sumatra. Can you see why? [*Source:* Courtesy of USGS.]

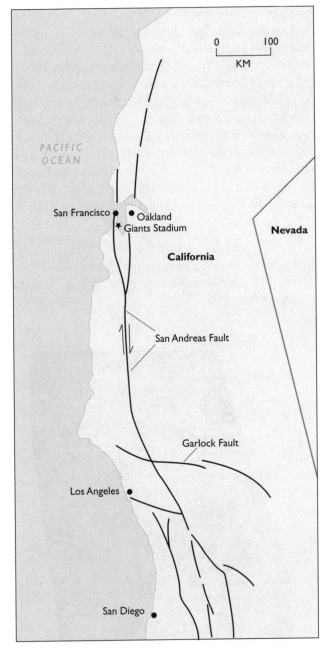

SAN ANDREAS FAULT

FIGURE 6-8 The San Andreas Fault System. Arrows indicate relative motion. Some have pointed out that eventually Los Angeles and San Francisco could be suburbs of each other, and the Giants and the Dodgers could again be cross-town rivals. Can you see why?

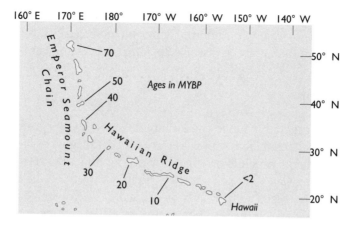

FIGURE 6-9 The Hawaiian Island Chain, showing ages in million years before present. The orientation of the islands shows the direction of movement of the Pacific Plate, which at present is towards the northwest. What was its direction of motion between 40 and 70 million years ago? [*Source:* Adapted from Dalrymple, G.B., et al., *American Scientist* 6 (1973): 294-308; Claque, D.A. and R.D. Jarrard, *Geol. Soc. Am. Bull.* 84 (1975): 1135-1154.]

■ Hot Spots, Plumes, and Island Chains

Plumes at abnormally "hot" sites under oceanic lithosphere produce volcanoes emitting mainly basaltic lava. A good example of such a volcano is Hawaii's Mauna Loa, the world's largest volcano.

In an ocean, a line of volcanic islands forms as the oceanic plate passes over the hot spot, as contrasted with the arc of islands at a subduction zone. The orientation of the island chain traces the direction of plate movement. The Hawaiian Plume is presently situated immediately southeast of the Big Island of Hawaii, and in addition to the active volcanoes on Hawaii, Kilauea and Mauna Loa, the plume is forming a new island called Loihi, presently entirely underwater.

Plumes may persist for tens of millions of years and apparently stay near the same place in the mantle. Figure 6-9 shows the Hawaiian island chain.

Hot spots also form along divergent boundaries. Iceland is formed by hot spot volcanism at the Mid-Ocean Ridge.

Summary: Oceanic Lithosphere

The oceanic lithosphere consists of, from base upward, mantle rock with a high magnesium (Mg) content, overlain by coarse textured rock of basaltic composition, in turn overlain by basalt. A layer of sediment, whose thickness and composition varies, lies on top. As new oceanic crust is created at mid-ocean ridges (MORs), the sediment layer tends to be thinnest and youngest there. With increasing distance from the MOR system, the layer of sediment tends to thicken, and the basal sediment becomes older. The composition of the sediment depends in part on water depth, which determines pressure, which in turn determines the solubility of some common oceanic minerals. At great water depths, for example, carbonate sediment, produced by floating, microscopic organisms living near the surface, is dissolved before it can accumulate on the bottom.

Marine Sediment

Marine sediment is generally youngest and thinnest nearest MOR segments. The present distribution of marine sediment is shown in Figure 6-10. Categories of marine sediment are evaporite sediment, which is produced by evaporating seawater (not shown on Figure 6-10), biogenic (*bio* = life and *genous* = origin) sediment produced by organisms and terrigenous (*terra* = land and *genous* = origin) sediment. We exclude here fragments of volcanic rock that form on the sea floor, as well as particles of meteorites.

■ Evaporites

Sediment produced by evaporating seawater is called evaporite sediment, or just evaporites. Seawater is approximately 3.5% (by weight) dissolved matter (discussed later here).

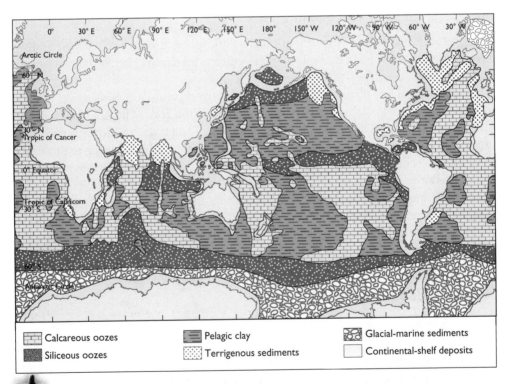

FIGURE 6-10 The Distribution and Nature of Deep-Sea Marine Sediment. Note the twin masses of "terrigenous" sediment at the northern end of the Indian Ocean. Given what you have learned about the formation of the Himalayas, what do you think the source of this terrigenous sediment to have been? [*Source:* Adapted from Davies, T.A., et al. *Chemical Oceanography*. Academic Press, 1976.]

Evaporating seawater causes different minerals to form as more and more seawater is progressively evaporated. In terms of volume, most evaporites are *halite*, or rock salt, as Na^+ and Cl^- make up more than 80% of dissolved matter. Other types of evaporites include gypsum (calcium sulfate) and sylvite (KCl).

■ Terrigenous Deposits

These materials (so named because they are fragments of pre-existing sediments formed on land) are carried into the ocean by rivers, wind, and glaciers. Terrigenous sediment may become altered and form new minerals after deposition on the sea floor by interaction with seawater, deep burial, or heat and gas from submarine volcanoes.

The most widespread terrigenous mineral in the oceans is clay. Clays are common in glacial and river sediments and may be blown great distances by winds. Clays and other sediments from the Sahara Desert can be carried all the way to the Caribbean region.

The most familiar terrigenous marine sediment is common beach sand, most of which is quartz. Quartz, composed solely of Si and O and very tightly bonded, is very hard to break down, and thus, after it forms, it can be carried long distances by rivers and eventually washed into the sea, where it is distributed along the shore by waves and currents. We have a great deal to say about beaches in Issue 3.

■ Biogenic Sediments

Organisms produce vast quantities of sediment as shell material. Most such material is calcium carbonate ($CaCO_3$), but some is glass-like material called silica. Deposits of silica (SiO_2nH_2O) may accumulate from the shells of microscopic marine organisms and silica-secreting organisms like sponges. Beaches far from a quartz sand source, like the islands of the Bahamas, are composed of soft carbonate sand. Much of this sand consists of broken and abraded pieces of nearshore coral. Organisms can produce shell material of a wide range of sizes, from many centimeters to fractions of a millimeter and even to sizes smaller than can be seen with a light microscope! The great chalk deposits of the English Channel and elsewhere are composed mainly of submicroscopic shells, made by minute organisms called coccolithophores.

The Sea Floor: MORs, Abyssal Plains, and Continental Margins

Figure 6-11 shows the major features of the ocean. Much of the sea floor is made up of a massive, globe-encircling seam-like feature called the MOR system. The MOR is where new oceanic bedrock is made; it tends to stand higher than other sea floor because it is hotter, and hot rock expands and occupies more volume than colder rock. Here is where molten material from the underlying mantle reaches the sea floor. The rising hot mush of molten material fills in cracks on the sea floor and causes it to expand laterally. Thus, in general, the youngest sea floor is along the axis of the MOR and sea floor bedrock gets older away from the MOR. Likewise, the bottom layer of marine sediment tends to get older away from the MOR, and this makes sense if you think about it.

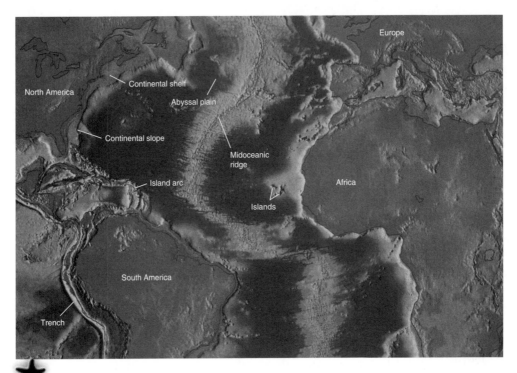

FIGURE 6-11 The Major Features of the Atlantic Ocean Floor. Based on the relative elevation of the abyssal plain and midocean ridge, which feature do you think is 'hotter'? Why?

The seafloor cools quickly away from the MOR. Next to the MOR are abyssal plains. Here the ocean bottom is deep because the underlying bedrock is cold. Original irregularities have been covered by layers of marine sediment over millions of years, and thus, at a global scale, the abyssal plains may look featureless. Far away from the MOR are trenches and adjacent island arcs. Here, subduction of old, cold oceanic bedrock is taking place. Carrying the slab of old ocean floor deep into the mantle causes it to heat and begin to melt. This partially melted oceanic bedrock forms new lava, erupting from time to time behind the trench to form arcs of volcanic islands like Japan and Indonesia. These features are labeled on Figure 6-6.

Now before we leave this section, take a moment and look back at the Hawaiian Islands on Figure 6-9. Their northern end, called the Emperor Seamounts, is oriented toward the north, but it makes a bend around Midway and turns southeasterly. Such abrupt changes in the orientation of island chains indicate times when the sea floor shifted its direction of spreading. If you know the age of the island in the chain on either side of the kink, you can tell when the shift occurred.

■ The Continental Shelf Complex

The continental shelf is that part of the continent flooded by the ocean. Its width ranges from a few kilometers to hundreds of kilometers. Its seaward border is the continental

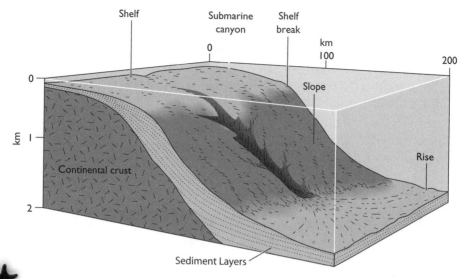

FIGURE 6-12 The Major Features of the Continental Margin. How do you think seismic waves could be used to locate the boundary between the continental crust and the sediment layers?

slope, where slopes steepen, plunging to the abyssal plain of the ocean floor. Figure 6-12 illustrates a continental margin typical of the eastern United States and Canada. The shelf is underlain by continental bedrock of two general types. The topmost layer is sediment eroded from the continental interior, underlain by older sedimentary rock, which is draped like a seaward-thickening blanket over older, mainly igneous and metamorphic rocks below.

Coasts

It can be one of life's most moving experiences to stand along the shore of a great ocean; however, earth scientists must classify such magical places to better understand them. One way to classify coasts is based on their plate tectonic setting. Active coasts occur at plate boundaries; passive coasts do not. Coastlines along the west coast of North America are active coasts, whereas those on the east coast are passive coasts.

The coast includes the ocean bottom adjacent to the shore. The shore is essentially the zone between the tides. The coast extends inland as far as the ocean influences landforms. According to the Coastal Zone Management Act of 1972, the coastal zone is the transition zone from land to the edge of the United States territorial sea, 3 miles from mean high water (average high tide), historically determined by the distance of a cannon shot. Coastal environments are often classified on the basis of the areas covered by tides. Intertidal areas lie between high and low tide. High-energy intertidal areas are beaches, dominated by sandy sediment, or rocky shores, with little sediment; low energy intertidal areas are tidal flats, dominated by muddy sediment. Subtidal areas are those below low tide, and supratidal land is above normal high tide but may be covered by very high tides.

Coral reefs are characteristic of many tropical coasts. They are massive structures built of calcium carbonate by colonial corals, calcareous algae, and a host of other organisms. Barrier reefs are coral reefs that parallel the land. As such, they are among the planet's greatest storehouses of carbon and play a critical role in the carbon cycle. Behind coral reefs, at the margin of the terrestrial environment, mangrove swamps are often found.

Coral reefs protect shorelines and harbor numerous organisms. They have, in addition, among the planet's highest levels of biodiversity. Reefs are important to the economies of many small island nations, especially in the Caribbean, by providing tourist income and fisheries for domestic consumption and export. Reefs are also major sources for the commercial tropical fish trade, in some cases with harmful consequences to the reef.

The energy for coastal erosion comes mainly from waves, which are disturbances of the sea surface. Most surface waves result from frictional effects of wind blowing across the water. Most of their erosive impact occurs during storms.

Development along coasts can be a severe environmental problem. Building along coasts can facilitate erosion and can introduce wastes into the coastal ocean. We have more to say on this subject in Issue 3. This promises to be an increasingly severe problem as the rate of sea level rise accelerates over coming decades, threatening hundreds of billions of dollars of investment.

What Causes Sea Level Change?

Sea level change is a phenomenon experienced by coastlines. Sea level has varied by more than 100 meters over the Earth's history. In addition to simple heating of seawater, which causes seawater to expand, sea level increases result from two processes.

- Eustatic processes change the absolute amount of liquid water within the ocean basins. The principal mechanism driving eustatic changes in sea level is the re-proportioning of water between H_2O liquid and ice because of changes in global climate. Our present climate is abnormally cold compared with the span of Earth history, but is warming at rates unprecedented in the past few million years. As our climate warms, sea level will undoubtedly rise.

- Isostatic processes change the underlying topography of the sea floor.

Isostatic changes can occur on local and regional scales, as in the rebound of crust after glacial melting, or in the slow subsidence of deltas and submarine fans at passive continental margins, or on global scales, as in periods of marked changes in sea floor spreading rates. As spreading rates increase, heat flow increases, and rocks of the ocean basins accordingly swell in volume. Expanding the sea floor in turn displaces water upward and outward onto the continent. It is important to stress that the rates at which isostatic processes occur vary widely. For example, sea level changes brought about by changes in spreading rates occur on million year time frames. Sea level rises and falls caused by changes in ice volumes can occur over hundreds of years, or even decades.

Insolation and the Earth's "Heat Budget"

The Earth's heat budget is a concept that relates the total solar energy received by the Earth (called insolation) to the amount of heat (1) reflected and (2) re-radiated by the Earth into space. Albedo is a term that expresses the reflectivity of a surface. Surfaces with high albedos have high reflectivity. Snow and ice are examples. Dark surfaces such as forests have low albedos.

As the geography of the Earth's surface changes, for example, as rock and forest are replaced by snow and ice, the amount of incoming solar radiation absorbed by the Earth declines. Currently, about a fifth of insolation is absorbed by the atmosphere. About a third is reflected back into space, and about half is absorbed by the surface and re-emitted as longer wave radiation, mainly infrared. The temperature of a body determines the wavelength of its emitted radiation: The cooler the body, the longer is the wavelength emitted. Very hot stars, for example, may emit extremely shortwave x-radiation. A cooler body such as the Earth emits mainly longer wavelength infrared and visible light. Gases in the atmosphere may be either transparent to, or opaque to, radiation that passes through it. If the gas is opaque to the radiation of a particular wavelength, the gas absorbs the radiation. A good example today is carbon dioxide in the atmosphere. As the carbon dioxide in the atmosphere increases, the ability of the atmosphere to absorb radiation increases, and the Earth becomes warmer: The heat budget thus changes. Many gases are opaque to the longer wave infrared radiation emitted by the Earth's surface. The two most important ones in the atmosphere are carbon dioxide and water vapor.

This is an example of the "greenhouse effect." Without a greenhouse effect, the Earth would be too cold to support life as we know it, but with a "runaway" greenhouse effect, the Earth could become hot enough to melt polar ice and inundate the margins of continents, as has happened many times in the Earth's past. We have much more to say on this topic in Issues 4–7.

Atmospheric Circulation

Winds result from the unequal heating of the Earth. The Earth receives most solar radiation at the equator and least at the poles. Thus, a heat gradient is set up, taking the form of a series of convection cells (Figure 6-13). As warm air rises and moves toward the poles, the winds are deflected by Coriolis force (explained on pages 75–76) to the right in the northern hemisphere and to the left in the southern. As the warmed air cools, it becomes heavier and sinks, again to be deflected by Coriolis. Cold, polar air flows along the surface toward the equator and is similarly deflected. Figure 6-13 shows the resultant global wind belts.

Seawater: Chemistry, Composition, and Structure

The volume of the planet's oceans can hardly be appreciated, even with the gigantic numbers we use. As a result, it can be difficult for people to believe that actions taken

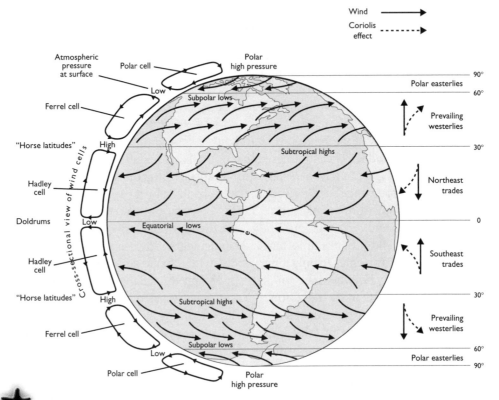

FIGURE 6-13 Convection, Atmospheric Pressure, Coriolis Force, and Resultant Global Wind Belts. Why is low atmospheric pressure typical of the equator, and high pressure typical of the poles? Hint: Which is denser:cold air or not air?

by humans can affect the entire oceans in any way. Sadly, however, pollutants created by humans have spread to every corner of the world's oceans, and many species of large marine mammals and fish are threatened with imminent extinction through overfishing and pollution, as we find in Issues 8, 10, 12, 13, and 18 (and others). Here we discuss the nature of the water in the oceans.

■ Origin of Seawater

How did the water in the world's oceans originate? Scientists have two hypotheses. First, bombardment of the Earth early in its history (perhaps 4 billion years ago) by icy comets (sometimes called icy snowballs because they are 75% ice) may explain the presence of Earth's surface water. A competing hypothesis argues that outgassing of water vapor from volcanoes, followed by condensation in the Earth's atmosphere, accounts for the planet's water. Which is correct? There is considerable evidence supporting both perspectives. Box 6-1 allows you to calculate whether the mantle of the Earth at present contains sufficient water to fill the world's oceans.

BOX 6-1

What Is the Original Source of the Water in the Oceans?

In this activity, you will determine whether the Earth's mantle is an adequate source of water for the world's oceans by performing the following calculations.

1. The total mass of rock in the mantle is estimated to be 4.5×10^{27} g. Assume that the mass of water locked in the mantle is 0.1% by weight (which is the upper limit for water content of peridotite, a common mantle rock).
   a. What is the mass of water stored in the mantle? (Hint: Multiply the mass by 0.1%, expressed as 0.001, or 1×10^{-3})
   b. Calculate the volume that this represents, assuming a density of 1 g/ml. Express your answer in liters. (Recall that 1 L contains 1000 ml.)

2. Now calculate the volume of water in the oceans. The average depth of the oceans is 3.8 km, and its surface area is 360×10^6 km^2. Express your answer in liters. (Hint: 1 km^3 contains 10^{12} liters)

3. Compare your answers to Questions 1 and 2. Does the mantle contain enough water to fill the world's oceans?

■ Composition and Chemistry of Seawater

Let us begin with pH. This is the measure of a solution's acidity, or the concentration of H^+ ions. The range of pH for various substances is shown in Figure 6-14. The typical pH of surface seawater is about 8.2, which is weakly alkaline. Because of all of the dissolved substances in seawater (discussed later here), the pH of seawater had previously been thought to be hard to change very much; however, scientists investigating climate change are becoming increasingly alarmed that higher levels of CO_2 in the atmosphere are leading to the acidification of seawater, as more and more CO_2 goes into solution in the oceans, especially in shallow water, where most organisms live. Adding CO_2 to seawater makes it more acid because the gas combines with water to form carbonic acid (H_2CO_3, which disassociates slightly to form CO_3^- and H^+). This could have potentially devastating consequences for marine organisms like colonial corals. The German Advisory Council on Climate Change reported in 2006 that over the period 1800–2005, the pH of the ocean surface water declined by about 0.11 units, from a preindustrial surface ocean pH of about 8.3. This corresponds to an increase in the concentration of H^+ ions (the measure of acidity) of about 30%. Furthermore, calculations indicate that if the atmospheric CO_2 concentration were to reach 650 ppm (parts per million) by the year 2100, this will cause a further decrease in the average pH by 0.2 units from present values.

Seawater is mainly H_2O with dissolved ions and gases. A measure of the amount of dissolved substances in seawater expresses its salinity. More specifically, salinity is the total weight, in grams, of all dissolved salts in one kilogram of seawater, usually expressed as

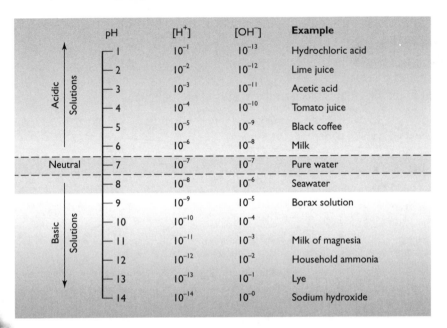

	pH	$[H^+]$	$[OH^-]$	**Example**
	1	10^{-1}	10^{-13}	Hydrochloric acid
	2	10^{-2}	10^{-12}	Lime juice
Acidic Solutions	3	10^{-3}	10^{-11}	Acetic acid
	4	10^{-4}	10^{-10}	Tomato juice
	5	10^{-5}	10^{-9}	Black coffee
	6	10^{-6}	10^{-8}	Milk
Neutral	7	10^{-7}	10^{-7}	Pure water
	8	10^{-8}	10^{-6}	Seawater
	9	10^{-9}	10^{-5}	Borax solution
	10	10^{-10}	10^{-4}	
Basic Solutions	11	10^{-11}	10^{-3}	Milk of magnesia
	12	10^{-12}	10^{-2}	Household ammonia
	13	10^{-13}	10^{-1}	Lye
	14	10^{-14}	10^{-0}	Sodium hydroxide

FIGURE 6-14 The Range of pH for Various Substances. How much more concentrated are H^+ ions in hydrochloric acid compared to lime juice?

parts per thousand (0/00). Table 6-2a shows the major elements dissolved in seawater. In addition, all naturally occurring elements are found to some degree in seawater. Table 6-2b shows representative examples and their concentrations. Gases, too, can readily enter seawater, often from mixing with the air by waves (a process termed bubble injection). Gases may also enter seawater from marine photosynthesis, when plants use CO_2 and give off O_2, and from volcanoes on the sea floor. Gas solubility is inversely proportional to salinity and temperature. Oxygen, for example, becomes less soluble as salinity and temperature rise.

Table 6-3 shows the average composition of gases and nutrients, essential to plant growth, in surface seawater. The amount of CO_2 and O_2 in seawater change with depth, due mainly to the effect of respiration, which depletes available oxygen between about 500–1500 m below the surface. Oxygen is added to surface water by photosynthesis and diffusion from the atmosphere and is added to deep water by oxygen-rich polar water flowing towards the equator along the bottom (see Figure 6-15a and 6-15b).

■ Cycling of Materials in the Ocean

Not only are gas and dissolved matter readily taken up by seawater, but material can also be lost. For example, a layer of oxygen-starved water (anoxic or hypoxic water) can occur if oxygen is depleted by aerobic respiration of organic matter, such as dead plant matter, sinking from surface waters. Such effects are brought about by the introduction of excessive nutrients from fertilizer and sewage and can form "dead zones," devoid of life, which we refer to in Issue 11.

TABLE 6-2A	Major and Minor Ions in Seawater of 35% Salinity	
Ion	**Chemical formula**	**Concentration (0/00) (ppt)***
Chloride	Cl^-	19.3 ⎫
Sodium	Na^+	10.6 ⎪
Sulfate	SO_4^{2-}	2.7 ⎪
Magnesium	Mg^{2+}	1.3 ⎬ Major
Calcium	Ca^{2+}	0.4 ⎪
Potassium	K^+	0.4 ⎪
Bicarbonate	HCO_3^-	0.1 ⎭
Bromide	Br^-	0.066 ⎫
Borate	$B(OH)_4^-$	0.027 ⎪
Strontium	Sr^{2+}	0.013 ⎬ Minor
Fluoride	F^-	0.001 ⎪
Silicate	SiO_4^{4-}	0.001 ⎭

Plus traces of other naturally occurring elements

*ppt = parts per thousand

TABLE 6-2B	Examples of trace elements in seawater
Trace Element	**Concentration (ppb)***
Lithium (Li)	170
Iodine (I)	60
Molybdenum (Mo)	10
Zinc (Zn)	10
Iron (Fe)	10
Aluminum (Al)	10
Copper (Cu)	3
Manganese (Mn)	2
Cobalt (Co)	0.1
Lead (Pb)	0.03
Mercury (Hg)	0.03
Gold (Au)	0.004

Examples of Trace Elements Dissolved in Seawater. About how much more concentrated in seawater are lead and mercury, compared to gold?

*ppb = parts per billion

TABLE 6-3

Quantities of gas in air and seawater

Gas	In Dry Air (%)	In Surface Ocean Water (%)
Nitrogen (N_2)	78.03	47.5
Oxygen (O2)	20.99	36.0
Carbon dioxide (CO2)	0.03	15.1
Argon (Ar), hydrogen (H2), neon (Ne), and helium (He)	0.95	1.4

Near-surface nutrient concentrations in seawater

Nutrient Element	Concentration (ppm)*
Phosphorus (P)	0.07
Nitrogen (N)	0.5
Silicon (Si)	3

*ppm = parts per million
Average Composition of Gasses and Nutrients in Surface Seawater. Why is silicon considered a nutrient?

Natural cycling of material in and out of seawater is described under the general heading of marine cycles. Let us take one example to illustrate cycles: carbon. CO_2 dissolves in seawater. Very cold water can dissolve more CO_2 than warm water. Some CO_2 combines with water molecules to form the weak acid H_2CO_3, called carbonic acid. Here is the formula for these reversible reactions:

$$CO_2 + H_2O <-> H_2CO_3 <-> HCO^{3-} + H^+$$

CO_2 may enter seawater from erupting volcanoes at the bottom of the sea, as you saw previously, but most CO_2 comes into the ocean via the atmosphere. CO_2 is readily dissolved in seawater, where it may be used by plants during photosynthesis: the most important reaction on Earth. The formula for photosynthesis is as follows:

$$CO_2 + H_2O + light\ (energy) -> CH_2O(n) + O_2$$

$CH_2O(n)$ is a carbohydrate (meaning essentially that the elements C, H, and O occur in a ratio of approximately 1:2:1), and O_2 is an oxygen molecule.

You saw previously here that there was abundant Ca in seawater (as Ca^{2+}). Animals such as corals and plants such as calcareous algae precipitate calcium to form a $CaCO_3$ "skeleton." This $CaCO_3$, now called limestone, represents stored or "fixed" CO_2—that is, CO_2 that has been removed from the ocean-air system. Similarly, microscopic diatoms use Si^{4+} in seawater to build a skeleton and thereby remove Si from seawater. At the same time, Si is brought into the ocean by rivers, replenishing the supply.

You are no doubt familiar with numerous marine animals that make use of $CaCO_3$ in shell construction. Limestone, the rock composed of calcium carbonate, is a valuable

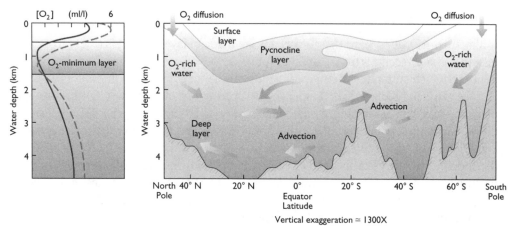

(A) VERTICAL O₂ PROFILES IN THE ATLANTIC OCEAN

(B) O₂-ADVECTION PATTERN IN THE ATLANTIC OCEAN

FIGURE 6-15 (A) The change in dissolved oxygen with depth in the ocean; (B) the means by which oxygen is added to deep and shallow, water. Oxygen-rich polar water enriches deep ocean water at depth. Oxygen solubility declines as temperature increases in the ocean. How could polar warming affect global deep ocean oxygen levels?

resource. India, for example, has limestone reserves of more than 179,000 million tonnes! Now consider again the formula of the calcite that makes up the limestone: $CaCO_3$. We can rewrite this as $CaO \cdot CO_2$. Nearly half of limestone by weight is CO_2. This rock was originally deposited in the ocean millions to hundreds of millions of years ago.

QUESTION 6-1. Pure limestone is 44% CO_2 by weight. Estimate the total weight of CO_2 "fixed" or locked up in India's limestone.

QUESTION 6-2. What would happen to the composition of the atmosphere if all of the limestone in India were dissolved into its constituents CaO and CO_2?

India's limestone reserves represent only a small portion of global limestone rock on land.

An important part of the carbon cycle is the exchange of CO_2 between the lithosphere, the atmosphere, living tissue, and the oceans. Some CO_2 is incorporated into the skeletons of animals and plants, and it may be "fixed" or removed from the system by various means. Ca and CO_2 were removed from seawater by organisms, and originally from the atmosphere, to make these rocks. As limestone is eroded and dissolved by rivers, the Ca and CO_2 (as HCO_3^-) re-enter river water and are carried back to the sea, from which they were removed ages ago. This is an example of the Carbon Cycle.

QUESTION 6-3. Suggest another example of the cycling of carbon at the Earth's surface. (Hint: Think of the composition of plants and animals. What happens when plant tissue decays?)

Scientists study several other key cycles involving the oceans, including the hydrologic (water) cycle, the nitrogen cycle, and the oxygen cycle.

■ Structure of Seawater

It may surprise you to learn that the oceans are made of layers, but it is true. Most of the heat the oceans receive is from sunlight, but volcanoes also provide heat from below. The depth to which sunlight penetrates depends on the turbidity (cloudiness) of the water and the wavelength of the light. Under the best conditions, most wavelengths of sunlight penetrate less than 500 meters which, when converted to heat, sets up a strong heat gradient. The boundary between the top, warmer waters and the colder dark waters below is marked by a strong temperature gradient found over most of the oceans called the thermocline. Figure 6-16 shows the thermocline. Thus, one cause of layering is uneven heating from above. Because salinity of most ocean water typically varies only slightly, temperature is the main determinant of seawater density over most of the ocean.

Another cause of layering is differences in salinity, which also cause differences in density—the more saline the water, the denser it is. Over most of the ocean, salinity is quite constant, averaging 35 parts per thousand, but in regions where a great deal of fresh water is added to the sea, from melting ice or a river for example, the salinity of surface water can be much lower. In regions where evaporation potential is high, such as the eastern Mediterranean, salinity can be significantly higher than the average. Figure 6-17 shows the average ocean salinity in August.

Density is important for many reasons. First, density differences determine the stability of seawater. If surface water becomes denser than underlying water through evaporation or cooling, the surface water becomes unstable and will sink until it reaches stability. Figure 6-18 shows examples of how density varies with depth. Denser water below less dense water produces a very stable layered situation. Density-driven circulation is responsible for distributing most dissolved substances in the oceans. Another way in which dissolved substances are distributed throughout the ocean is via circulation of unstable waters above active sea-floor volcanoes. Finally, the

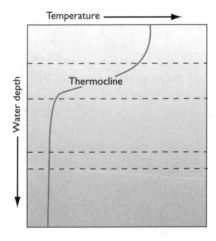

FIGURE 6-16 The thermocline. Why is ocean water so consistently cold below the thermocline?

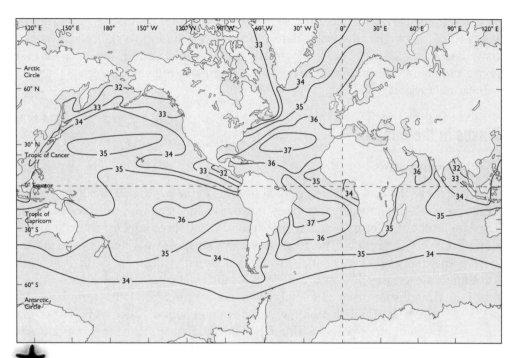

★ FIGURE 6-17 Average Sea-Surface Salinity in August. What is the range in sea-surface salinity? "Average" ocean salinity is often given as 35 parts per thousand (0/00). Is this justified, based on the figure? Why or why not? [*Source:* Adapted from Sverdrup, H.U., et al., *The Oceans.* Prentice-Hall, 1942.]

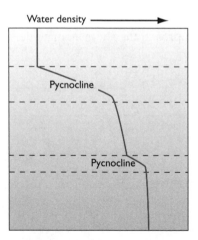

★ FIGURE 6-18 Variation of density with depth. The Pycnocline is a layer where density changes. Why do you think density is consistently low at the surface?

distribution of biologically important O_2 and CO_2 is controlled by animal respiration, photosynthesis and oxidation by microbes.

Second, density determines the speed of sound within the oceans. Other things being equal, the higher the water density the faster sound moves. Oceanographers have discovered discrete layers within the oceans where sound travels faster than above and below. This information is of great interest to military strategists planning submarine warfare.

Waves in the Ocean

■ Wind-Driven Waves

When we think of waves on water, most of us think of the waves that break on a beach. These are wind-driven waves, and they transfer energy from the wind to the water. They are responsible for moving material around at the shore, and for coastal erosion, so they are very important. The greatest energy transport by water waves is during storms. They move water only in the directions indicated by Figure 6-19. In deep water, wave motion decreases downward to a depth called wave base, below which there is no effect. Thus, a submarine or a whale can avoid the most powerful storm by diving below the wave base. Along the coast, waves encounter bottom above wave base so that the waves begin to move particles along the bottom, perhaps causing significant erosion or movement of coastal sediment. The depth to which waves affect water in coastal regions can be estimated using the formula $D = L/2$. L, the wavelength, is the distance between two equal points on a wave as seen on Figure 6-19. When the water depth drops below about half of the wavelength, the friction of the wave dragging across the bottom begins to affect the entire wave. Wave height increases, and wavelength decreases. The wave begins to break when the top, moving faster than that part of the wave near the bottom, spills over. The energy thus released is focused on the beach, and this is a major cause of erosion. If waves approach the shore at an angle, which is the usual case, the

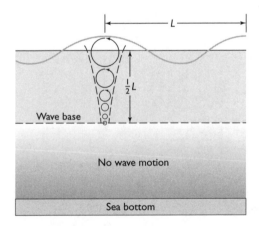

FIGURE 6-19 Movement and Characteristics of Wind-Driven Waves. What does the term 1/2L represent? What is its significance?

waves move material along the shore. Beach sands have often been called sand rivers because of this longshore transport caused by waves.

The size of a wave depends on the wind speed, the duration that the wind blows, and the distance over open water that the wind has blown (the wind's fetch). Sustained high-speed winds combined with a long fetch produces large waves with great potential energy. Waves represent wind energy originally powered by the sun, transferred from the atmosphere to the water.

Other types of waves found in the ocean besides wind waves are tides, seismic waves, and sound waves.

Tides

Tides are waves with very long periods. The period, T, = 1/frequency. Frequency in turn is the number of waves that pass a point per unit of time, seconds, minutes, or even hours in the case of tides. Tides are generated by the interactions of the gravitational pull of the Sun, Earth, and Moon with the Earth's rotation. A rotating body tends to spin material away from its surface, much like a dog spins water away when it shakes its head. Water at the Earth's surface is affected by tides of course, but it may surprise you to learn that rocks, and even ice sheets, are affected by tides as well.

Figure 6-20 shows how tides affect the oceans. To understand how tides cause a point to experience rising and falling seas, imagine that the Earth is turning under the bulge, as the bulge only moves with the Moon and Sun. There are two "bulges" in the Earth–Moon system. This is because the Earth–Moon system actually revolves around the center of its masses which, because the Earth is so much more massive lies about 1,500 km below the Earth's surface and about 5,000 km from the Earth's center. Also, the tidal bulges are bigger toward the moon than toward the Sun. Why is this? The answer can be understood from Newton's gravitational formula. He expressed gravity this way: F, the force of gravity, is proportional to $(M \times m)/d^2$, where F is gravitational force, d is the distance between the bodies, and M and m are the bodies' masses. We read it this way: "The gravitational force (F) between two bodies is directly proportional to the product of their masses $(M \times m)$ and inversely proportional to the square of the distance (d) between their centers." Even though the Sun is much greater in mass than the Moon or the Earth, it is so much farther away that the d^2 component of the equation is most important. Now to convert any proportionality to an equation, we have to introduce a constant (K). In Newton's equation, g is the gravitational constant. Thus, the rewritten equation is $F = g \times m \times M/d^2$.

You can see from Figure 6-20 that the tidal force and resultant effect are greatest along a line connecting the centers of M and m. Tide strength at any given place is affected by many variables, such as the morphology of the coast. The shape of the coast helps to determine the difference between high and low tide. This is most spectacularly seen in North America in the Bay of Fundy (Figure 6-21). Such sites, although rare, offer interesting possibilities for renewable energy development, if appropriate environmental safeguards can be put in place.

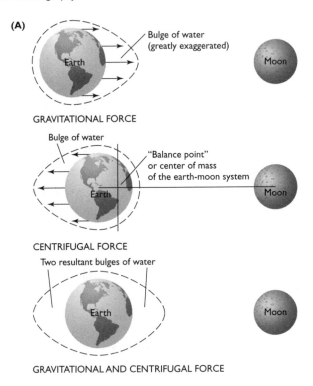

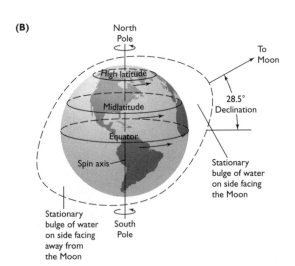

FIGURE 6-20 Ocean Tides. (A) the balance between gravitational force and centrifugal force produces tidal bulges. The bulges should be strictly viewed as cones. Can you see why? (B) Fixed points on the Earth's surface rotate into and out of fixed tidal bulges, or cones. Which way is the earth rotating?

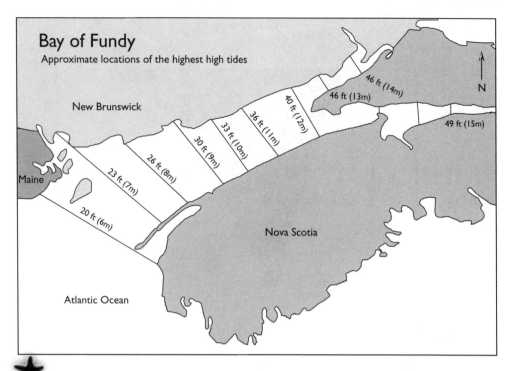

FIGURE 6-21 Tides in the Bay of Fundy. What is the maximum tidal range in the Bay of Fundy? [*Source:* Courtesy of NOAA.]

■ Seismic Waves

Seismic waves are shock waves, most commonly produced by earthquakes. They can also be produced by submarine landslides and by meteorite impacts. One of the most destructive effects of earthquakes are seismic sea waves, called by their Japanese name, *tsunami.* The December 2004 earthquakes off the coast of Indonesia produced the largest tsunami ever directly recorded by humans. Many coastal areas, especially in the Pacific and Indian Oceans, are susceptible to tsunamis.

■ Light

The actual depth to which light penetrates depends on suspended matter, turbulence, and wavelength, as shown in Figure 6-22A and 6-22B. Light, as humans perceive it, penetrates only slightly into the oceans' average depth of 3,800 meters. Marine creatures that live at great depths, however, often have extremely sensitive eyes. Much of the light received at the ocean surface is reflected, especially if the light strikes the surface at an angle. In the tropics, however, the sun is usually directly overhead, and thus, more light is absorbed, helping to warm the tropical ocean.

At the poles, where the sun angle is always low at best, most light energy is reflected; however, ice-covered water reflects far more of the light's energy than does ice-free polar

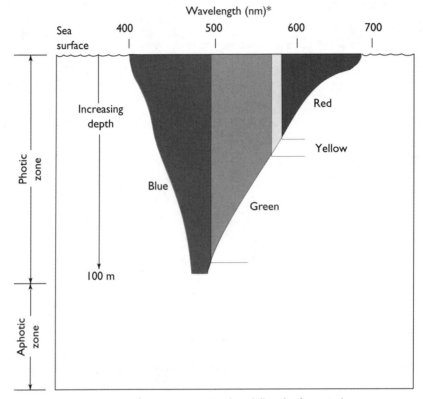

(A) LIGHT ABSORPTION IN THE OPEN OCEAN

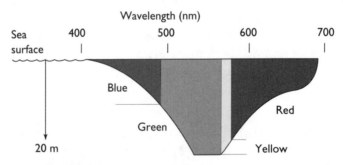

(B) LIGHT ABSORPTION IN NEARSHORE WATERS

FIGURE 6-22 (A) Depth to Which Light Penetrates in the Ocean. What is the depth of the "photic zone" in the open ocean? (B) Sediment in coastal waters reflects and diffuses light, so it penetrates much less deeply here. What is the longest wavelength of visible light? What color does it represent? [*Source:* Adapted from Levin, G.S., *Oceanus* 23 (1980): 19-26.]

ocean water. Albedo is, as we mentioned previously, a measure of the reflectivity of a surface: Those surfaces that reflect most light have highest albedos.

If Earth were covered in ice, its albedo would be about 0.84, meaning it would reflect 84% of incoming sunlight. For comparison, if the planet were completely covered by forest, its albedo would be about 0.14. Satellite measurements indicate Earth's average albedo to be about 0.30. Scientists have expressed concern that global warming is rapidly melting summer polar ice, which could potentially lower the polar region's albedo, thus altering the entire heat budget (the balance between incoming and outgoing energy) of the Earth, with poorly understood consequences.

■ Sound Waves

The speed of sound increases with the density of the medium through which it is passing. Thus, sound travels faster in water than in air. As seawater density changes, sound velocity changes too. Density depends on temperature and salinity. As temperature goes up, density goes down, and as salinity goes up, density goes up.

The intensity of sound is measured in decibels. Figure 6-23 shows this scale. The scale is logarithmic: That is, an increase in sound intensity from 10 to 20 decibels indicates a doubling of sound intensity. This is very important in Issue 15, the impact of high-intensity sounds on marine mammals.

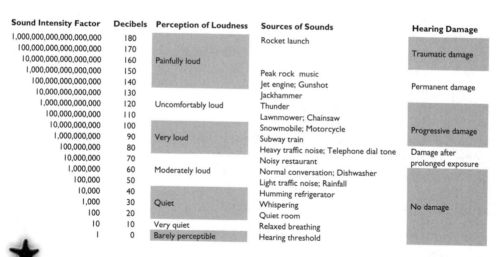

FIGURE 6-23 The Decibel Scale. In Dorothy Sayers' mystery novel "The Nine Tailors", a man dies because he is trapped in a church bell tower during a Christmas bell ringing. Estimate the decibel sound level he was exposed to that caused his death. [Adapted from Jonathan Turk and Amos Turk, *Physical Science with Environmental and Other Practical Applications*, Third edition, Saunders College Publishing, 1987.]

. . . .

Marine Circulation

■ Thermohaline Circulation, Gyres, and Currents

We described the origin of wind belts earlier, on page 45. Ocean water circulates, arising out of (1) the Earth's rotation, (2) density differences in ocean water, (3) the positions of continents, and (4) the Earth's wind belts. These values differ and change, albeit at different rates and with much variation. For example, the rate at which continents move is on the order of a few centimeters a year. The Earth wobbles on its axis over a cycle of tens of thousands of years, and wind belts shift with the seasons.

Marine scientists study ocean circulation in three dimensions but it is easiest to visualize circulation as having a horizontal and a vertical component. Figure 6-24 shows the horizontal component at the surface. Vertical marine circulation driven by heat and salinity differences is called thermo (= heat)-haline (= salt) circulation. Here is how it works. Warm saline water is commonly carried by ocean currents from the tropics to polar regions, where it is intensely chilled. This now cold, salty water may become denser than the water below it and so may sink. This sinking sets in motion other masses of ocean water throughout an ocean basin and is the primary driver of vertical circulation.

Figure 6-24 shows that each ocean basin contains a gyre, consisting of surface currents outlining a roughly circular pattern of movement. Where there are no continents in the way, a circum-global current persists. Currents moving from the equator towards the poles carry warm water. Those moving from the poles towards the equator carry colder water. In this way, the surplus heat generated by the high sun angle and great heat absorption at the equator is distributed around the Earth.

Look especially at the North Atlantic in Figure 6-24. We could spend this entire book describing the fascinating nature of this ocean basin and its circulation but must content ourselves with a few brief points. The Gulf Stream carries vast amounts of heat from the Caribbean northwards where prevailing westerly winds in mid-latitudes (roughly 40–60 degrees) help warm Western Europe. Largely because of this, palm trees grow in protected places along the west coast of Ireland! If the Gulf Stream were to stop or even slow, the climate of Europe would drastically change, and this is what could happen with global warming as we discuss in Issue 6.

■ Ekman Transport and the Global Garbage Dump

Now we explain a feature of gyres called Ekman transport, which results from the Coriolis effect (Figure 6-25).

The Coriolis Effect explains the apparent deflection of a fluid, in our case, water, resulting from differences in the speed of the Earth's rotation with latitude. The Coriolis effect is defined as an apparent deflection of moving objects to the right in the northern hemisphere and to the left in the southern hemisphere. A simple example will facilitate understanding the Coriolis effect. If you jump while standing in the aisle of a jet moving at 1,000 km/hr (600 mi/hr), you do not land in the aisle several rows back, or

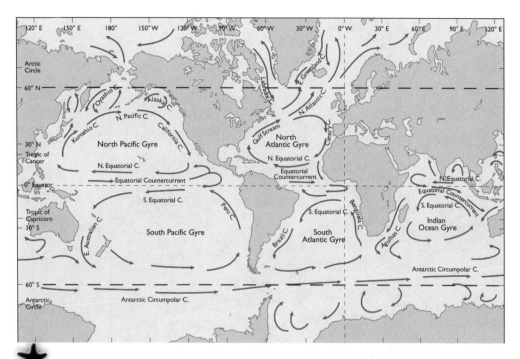

FIGURE 6-24 Surface Ocean Circulation. If you were planning a sailing ship's route from England to its North American colonies in 1770 and back again, and had access to a version of this map, what route would you take?

worse, slam into the rear of the plane. Why not? Although you perceive that you, your fellow passengers, and all of the jet's interior fixtures are stationary, all of these are in fact moving along in the same direction and at the same speed as the jet. Now consider the rotating Earth, moving from west to east (counterclockwise rotation). Imagine a jet plane flying from, for example, Hawaii, toward the equator. Where the plane takes off, the Earth's surface has a rotational velocity determined by the circumference at that particular latitude (in miles or km) divided by 24 hours, the time it takes the Earth to make one rotation. Say the circumference at Honolulu is 32,000 km (20,000 miles). The rotational velocity would thus be 32,000/24 or 1,330 km/hr (or about 800 mi/h). Thus, just as you were moving at 1,000 km/hr (the same speed as the jet) in the preceding example, this plane leaves Honolulu with a rotational velocity of 1,400 km/h (870 mi/hr, the same velocity as the Earth) in addition to its air speed. When the plane reaches the equator, however, the circumference there is 40,300 km (25,000 mi). Thus, the rotational velocity is 40,300/24, or about 1,700 km/hr (1,100 mi/hour); therefore, to an observer in the plane, the plane would seem (hence, the word *apparent* in the previous definition) to have been deflected to the right. In other words, the plane would be more and more left behind as it approaches regions of the Earth with higher and higher rotational velocities, as the plane inherited its rotational velocity from the latitude where it took off.

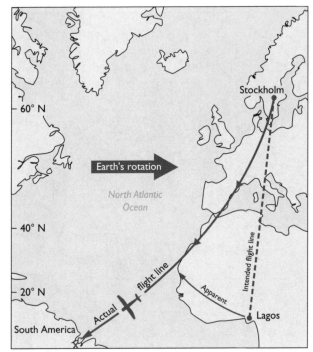

(A) CONSEQUENCES OF CORIOLIS DEFLECTION

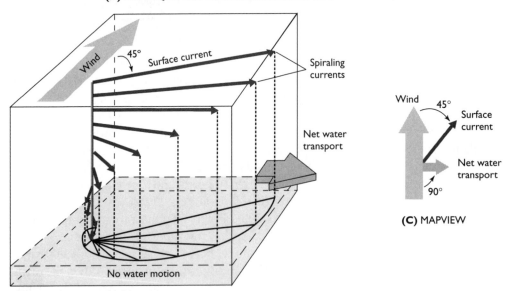

(B) EKMAN SPIRAL IN THE NORTHERN HEMISPHERE

(C) MAPVIEW

FIGURE 6-25 The Coriolis Effect and Ekman Transport. 6-25(A) shows the route of a jetliner from Stockholm to Lagos. How does the Earth's rotational velocity change from Stockholm to Lagos? 6-25(B) shows the direction of net water transport arising out of Ekman Transport. Look back at Fig. 6-24. The North Pacific Gyre contains among the planet's highest concentration of floating plastic debris. Based on Ekman Transport, suggest a source of most of this plastic.

Ocean water masses behave in a similar manner. Now study Figure 6-24, and look at the North Pacific Gyre. Coriolis diverts the water column in the currents comprising the gyre, with decreasing effect downwards as seen in Figure 6-25. This spiral-like deflection of ocean water layers is called the Ekman Spiral, and the resulting transport of water 90 degrees to the direction of the surface current is called Ekman Transport. The net effect of Ekman Transport is that the water is "pushed" into the center of the gyre. This area in the North Pacific has one of the greatest collections of floating waste, mainly plastics, on the planet, all brought into the area by winds and currents, and concentrated in the center of the gyre by Ekman Transport.

■ Upwelling

Upwelling occurs when wind persistently blows in one direction over the coastal ocean, relentlessly pushing the surface ocean water away, aided by Coriolis deflection, to be replaced by water from below. Surface water may become depleted in nutrients (nitrogen, phosphorus, silicon, etc.) by the activity of photosynthesizers. If nutrients are not replaced, photosynthesis will eventually stop, threatening the food web. It is no accident that upwelling zones are zones of high productivity in the oceans, as here, nutrients that are depleted by plants at the surface are replaced by upwelling, nutrient-rich waters from below. Figure 6-26 shows how this operates. You can see that the wind blows from the

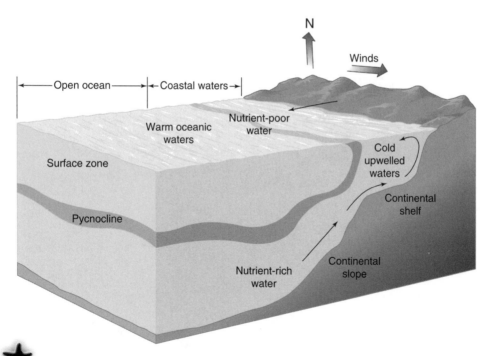

FIGURE 6-26 Upwelling. How does upwelling result in nutrient enrichment of coastal waters? How could warming of seawater by climate change contribute to "dead zones," devoid of life, in coastal waters?

north, deflecting the surface water to the west (or right: remember the Coriolis effect!). The deflected surface water is replaced by water from below, replenishing the nutrients that may have been depleted from the surface water.

■ Dead Zones

Dead zones occur at depth when waters become depleted of oxygen, forming hypoxic (low-oxygen) or anoxic (no oxygen) conditions. These can be very destructive to marine life as you may well imagine and can play havoc with marine fisheries, for example. Dead zones are apparently becoming more and more common, as pollution in the form of fertilizer- or sewage-enriched river water enters the ocean in greater and greater quantities. The effects are simple and deadly. The nutrients in the river water are carried into the ocean where they cause a spike in the growth of surface plants, mainly algae. These plants eventually use all of the nutrients and die. Their bodies decompose, using oxygen in the process. Dissolved oxygen is essential to animal life, and thus, as it diminishes and even disappears, animals either die or migrate away. A persistent dead zone exists off the mouth of the Mississippi in the Gulf of Mexico. In Issue 11 we discuss the nature and impact of the Gulf of Mexico and Oregon dead zones.

Essentials of Marine Ecology

What Is Ecology?

The word *ecology* comes from two Greek words (*oikos* and *logos*) meaning "household" and "knowledge of." It is the systematic study of the interactions among organisms and between organisms and their environment and is based on the recognition of ecosystems. An ecosystem is an area or region that has distinctive life and environmental conditions. From Chapter 6 you saw that one fundamental way to divide the oceans is into lighted and unlighted areas. We call these photic and aphotic zones. Plants require light for photosynthesis. Any region in the oceans that receives insufficient light for photosynthesis will have no plants and will be part of the aphotic zone.

One simple way to divide the ocean is into the shallow-water, photic habitat and the deeper water, aphotic habitat. The boundary between these two zones, that is, the depth to which sufficient light penetrates for photosynthesis, depends on conditions in the oceans. In coastal regions with higher turbulence and more suspended sediment, the photic zone is relatively shallow. In the open ocean, it can be more than 100 meters (300 feet) deep. The photic zone is a key concept in the ocean because plants depend on sunlight, and most creatures in the oceans depend ultimately on photosynthesis for sustenance.

Important Marine Habitats

Ecology is the study of relationships. The interactions between organisms occur in environments called habitats, that is, where they live. Habitats are characterized in turn by a set of parameters. Ninety-seven percent of open ocean environments are below 200 meters depth. Ninety-two percent of bottom environments are below 200 meters depth.

In no particular order, here is a list of important marine habitats. We begin with coastal habitats.

■ Coastal Habitats

Coral reef. The coral reef is a very complex tropical shallow water environment based on carbonate-secreting, colonial coral animals. Coral reefs, a type of hard bottom community, have the most different kinds of organisms (have the highest diversity) of any marine habitat. Reefs are under severe threat from urbanization, overpopulation, pollution, tourism, and overfishing. Already perhaps a third of Earth's coral reefs have been degraded and some perhaps irretrievably lost (see Issue 16 for more information on coral reefs).

Mangrove swamp/forest. Mangroves are tropical trees that can tolerate salt. Mangrove swamps are important nursery grounds for larval and juvenile fish and other organisms and protect the coast from tropical storms and tsunamis. Mangrove swamps are under grave threat from coastal development for housing and aquaculture. Issue 17 deals with mangroves.

Salt marsh. A salt marsh is essentially a low-energy coastal environment roughly equivalent to a mangrove swamp outside the tropics. Salt marshes are very diverse, productive environments and are under threat from pollution, subsidence, oil and gas mining, and development.

Experts believe the destruction of New Orleans by Hurricane Katrina could have been mitigated if the offshore marshes had not been degraded.

Estuary. An estuary consists of variably salty (brackish) water, where fresh water and salt water meet and intermix. Estuaries (e.g., Chesapeake Bay) are often found in embayments at the mouths of rivers. Even though they are less diverse than, for example, coral reefs, they are very important economically and are under threat from pollution, development, invasive species, and overfishing. For example, the native oysters in Chesapeake Bay, so numerous as to have awed early English settlers, are almost completely gone (see Issue 10 for more discussion about estuaries).

Other habitats. Other important coastal habitats include the following:

- *Kelp beds* (also called kelp forests). These are large offshore areas of fast-growing brown algae that grow in clear, primarily shallow water. Kelp beds are very productive environments that provide important habitat for marine organisms, including endangered marine mammals such as sea otters.

- *Seagrass beds.* These are home to submerged aquatic vegetation (SAV). Seagrass beds, composed primarily of marine flowering plants, are diverse systems found in shallow waters. Think of seagrass beds as ecosystem engineers: They create habitat by slowing water flow and increasing sedimentation. Sea grasses are indicators of ecosystem health and are disappearing globally because of human disturbances, for example, nutrient pollution in Chesapeake Bay.

- *Live bottom habitat.* Near-shore live bottoms are benthic (bottom) ecosystems on the continental shelf characterized by extensive assemblages of attached algae and sponges, soft corals, and other invertebrate animals (those without backbones), often growing on rocky outcroppings. Coral reefs (see above) are a type of live bottom habitat.

- *Beaches.* Beaches are high energy, dynamic, unstable environments with coarse sediments, all of which represent harsh conditions for organisms. Despite the widespread distribution and popularity of sandy beaches, surprisingly little is known about the ecology of them. These systems are threatened by development, which destroys habitat and alters the dynamics of sediment movement in the system. Tide pools—small, isolated, ephemeral pockets of water—frequently form in the upper intertidal zone along rocky shores. These may harbor an extremely rich and diverse biota.

■ Open Ocean Habitats

Photic zone. This zone includes water to the depth at which light is no longer sufficient for photosynthesis. At the top of the photic zone is the planktonic environment. This is a zone of floating and weakly swimming organisms, containing phytoplankton ("plant" plankton, primarily microscopic organisms known as diatoms and dinoflagellates that use energy from the sun to manufacture food) and zooplankton ("animal" plankton). The zooplankton consists of plant eaters (herbivores) and predators (carnivores). Some zooplankton spend their entire life in the plankton (holoplankton), whereas others spend only their egg and/ or larval stages (meroplankton).

Benthic. Benthic means bottom. The bottom environment can be photic or aphotic, but based on the ocean's average depth of 3800 meters, you can see that most benthic environments are aphotic. One can divide this environment into shelf benthic and deep benthic. The shelf benthic environment, closer to shore, ranges from the shallow subtidal to the edge of the continental shelf in more than 100 meters (300 feet) of water. Sediment type can include terrigenous coarse to fine, carbonate, and even glaciomarine (gravel to clay) debris. Carbonate sediment is common in tropical latitudes, whereas glaciomarine sediment is presently being deposited in some polar latitudes.

Deep benthic. The deep ocean bottom is devoid of light as it is typically several thousand meters below the surface of the sea. It has no photosynthesizers, but organisms that filter feed and scavenge can survive here, as a surprisingly steady "rain" of organic debris may be deposited from planktonic organisms in shallow water far above. The weight of water in the ocean means that as depth increases pressure increases significantly. For example, the entire column of air in the atmosphere at any given site at sea level only weighs about 14.7 pounds per square inch or 1 metric ton per square meter (spelled *tonne* after the French, who after all invented it). This value, 1 tonne/M^2 is called one atmosphere of pressure. In the oceans, however, the greater mass of seawater (water + dissolved matter) means that for every 10 meters increase in depth the pressure increases by one atmosphere. Organisms in deep water or on the deep bottom must thus be able to tolerate high pressures, which compress gas-filled spaces (like swim bladders in fish) and even alter the structure of some molecules (like proteins).

A recently discovered (1970s) deep-bottom ecosystem is that found around mid-ocean ridges, known as deep-sea hydrothermal vent communities (Figure 7-1). Volcanic gas and water are vented at high temperature (up to several hundred degrees C) and contains much dissolved matter, including cations such as iron (Fe^{2+}), manganese (Mn^{2+}), sulfur (S^+), and many others, but no free (uncombined) oxygen gas. Bacteria and other "primitive" organisms carry out chemical reactions to extract energy from these ions and gases and in turn are ingested by other animals. No photosynthesizers live in these environments of course.

Open Ocean Environments and Nutrients

The open ocean water, far from shore, is often called a "biodesert" because of the lack of nutrients (N, P, Si) to support photosynthesizers. Nutrients are usually brought into the sea by rivers, and thus, ocean sites far from rivers may contain few nutrients; however, nutrients can enter the open ocean by airborne dust, and by air deposition from fossil-fueled power plants, perhaps thousands of kilometers from the plants themselves. For example, pollution from Asian power plants is being recorded along the West Coast of the Americas. Because the amount of nutrients in the water controls photosynthesis, a lack of nutrients means that animals that depend on plants cannot survive, even in

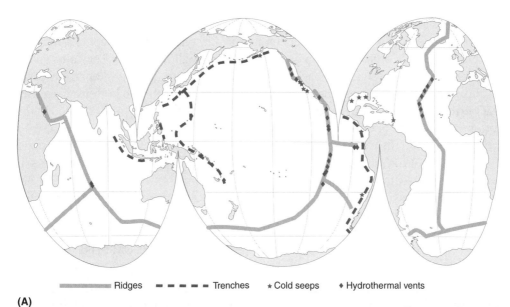

Ridges ▬▬▬   Trenches ▬ ▬ ▬   ★ Cold seeps   ◆ Hydrothermal vents

(A)

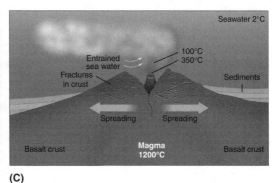

(B)   **(C)**

FIGURE 7-1 Hydrothermal Vent Communities. (A) The approximate location of communities. How is the location of these communities related to plate boundaries? (B) Tube worms and clams, important members of the vent communities. Tube worms were not known to exist before being discovered in the 1970s. (C) A cross-section of a Midocean ridge axis, showing the conditions necessary to maintain vent chimneys. Given sea floor spreading, are individual vent communities permanent? Why or why not?

the photic zone. The growth of phytoplankton is usually strongly seasonal because of limited nutrients and variations in the availability of sunlight.

Marine Organisms

A detailed classification of life is beyond the scope of this book. Students who have access to the Internet have a wealth of information at their fingertips. Here we briefly introduce the startling abundance and diversity of life in the oceans and focus on the organisms that we will feature in the issues that follow.

We can start to understand marine animals by describing their "lifestyles," or where and how they live. There are three general marine lifestyles: planktonic, nektonic, and benthic. Planktonic organisms float or swim feebly at shallow depths. They are usually microscopic, but may be as large as oceanic sunfish or even basking sharks. Nektonic organisms are free swimmers. Benthic organisms live on, or in, the bottom. Creatures that burrow into the bottom are called infauna, and those that live on or attached to the bottom are epifauna or epiflora.

To understand the variations among benthic organisms, we also need to understand something about the specific bottom environment. Is it composed of hard rock or soft sediment? Is the sediment muddy or sandy? Is the bottom photic or aphotic? Finally, are currents quiet or vigorous? These specific environmental parameters will determine the nature of the organisms that inhabit any particular bottom.

After we have assessed the differing "lifestyles" of organisms, we need to consider how they obtain food. There are only a few key ways organisms can make a living in the oceans.

1. Organisms can make their own food (primary producers). These include photo-synthesizers; plants (and plant-like organisms) that make their own food using CO_2, nutrients (N, P), and sunlight for energy; and chemosynthesizers, bacteria, and other primitive organisms especially on the sea floor above and around hydrothermal vents. These organisms use heat and elements dissolved in volcanic gas to make food. Organisms that make their own food are the basis of all life in the oceans.

2. Organisms (called primary consumers) can eat primary producers. Organisms that eat primary producers are also called herbivores. They are essential in the oceans as they transfer energy to carnivores and other larger animals by being eaten. A few marine mammals (manatees and dugongs), fish, and many invertebrates are primary consumers.

3. Organisms can eat other animals. Organisms of all kinds and sizes make a living by eating other organisms. They are called carnivores, or secondary and higher order consumers. They may actively chase others or may lay in wait and pounce or they may filter prey out of the water (filter feeders). Filter feeders can be tiny animals a fraction of a centimeter long or as gigantic as whale sharks, 10 meters long!

4. Organisms can eat dead matter. These organisms may be deposit feeders, crawling through sediment and ingesting it, suspension feeders that depend on debris from above, or surface scavengers. They are essential to recycling material in the ocean. In nature, waste is a symbol of inefficiency. In natural systems, there is no waste.

■ Classification of Organisms

Biologists classify organisms on the basis of the number of shared similarities. Those with the greatest number and that can interbreed to produce fertile offspring are members of the same species. Species are grouped into genera (singular = genus), genera into families,

families into orders, orders into classes, and classes into phyla (singular = phylum) and divisions.[1]

Classification schemes, like organisms, evolve as scientists use techniques of molecular genetics to understand more fully the relationships between organisms. This has resulted in a transition from a system of classification that groups closely related phyla and divisions into five kingdoms, the most inclusive classification category, to one in which the overarching system is three domains. We present a brief overview of both approaches.

The five kingdoms, which are separated based largely on different modes of nutrition and cellular organization, are as follows:

- Kingdom Monera. Single-celled organisms with a prokaryotic cell organization (Figure 7-2). Prokaryotic (pro = before, karyon = kernel, referring to the nucleus) organisms lack a well-organized cell nucleus and organelles surrounded by membranes. The remaining four kingdoms are eukaryotic (eu = true), meaning that they have a true nucleus and organelles such as mitochondria and chloroplasts that have a membrane. Monerans include bacteria and Cyanobacteria.[2]

- Kingdom Protista. A very diverse group of single-celled organisms with a cell nucleus. Examples are foraminifera, diatoms, and dinoflagellates (Figure 7-2). The latter two groups are plant-like and are called algae.[3] Protists are major components of plankton.

- Kingdom Plantae. Marine plants, either microscopic or visible, and either floating or attached. These include green, red, and brown algae, and sea grasses.

- Kingdom Fungi. Multicellular, filamentous organisms that absorb food (rather than ingest it); though not very abundant, marine fungi are important decomposers of organic matter.

- Kingdom Animalia. Multicellular animals, including vertebrates and invertebrates.

The alternative three domains system consists of the following:

- Domain Archaea. Prokaryotic organisms that inhabit extreme environments like deep-sea hydrothermal vents and hypersaline waters. Members of this domain differ from other prokaryotes in the composition of their genetic material and the structure of membranes and flagella.

- Domain Bacteria: Bacteria, which include the remaining prokaryotes.

- Domain Eukarya: All of organisms with eukaryotic cellular organization.

Here are some important types of marine organisms.

---

1. Traditionally, plants and plant-like organisms have been grouped into "divisions," which are the equivalent of phyla.
2. Cyanobacteria are unfortunately commonly known as blue-green algae. All algae are classified as eukaryotic and cyanobacteria are prokaryotic. The term "blue-green algae" is firmly entrenched in the taxonomic lexicon, and thus persists.
3. The term "algae" has been commonly used to include plant-like members of the kingdoms Protista and Plantae.

Bacteria and archaea are very important in the oceans, mainly as decomposers of waste and tissue, enabling these to be reused by other organisms. They are numerous around volcanic vents in the deep ocean, where they convert chemical energy into tissue and serve as the basis for the food web at those sites.

Many protists are tiny, free-floating forms making up planktonic phytoplankton. In most ecosystems, they are the critical basis of the oceanic food web. Some protists live

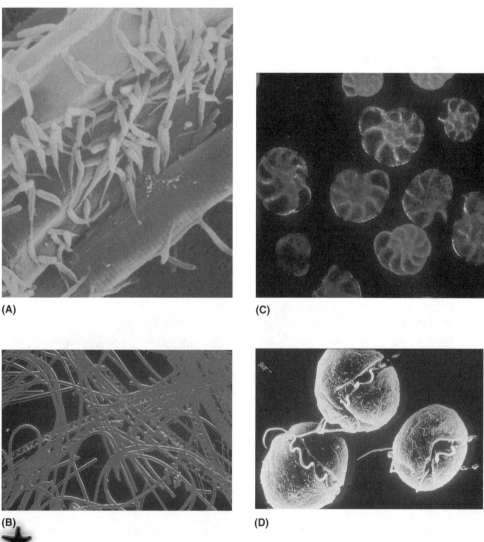

(A)

(C)

(B)

(D)

FIGURE 7-2 Examples of Monera and Protists. (A) Marine bacteria, (B) Blue-green algae. Blue-green alga, at 0.002 inches in length, are about ten times larger than the bacteria. (C) Foraminiferans, (D) Dinoflagellates.

in tissues of many marine animals, providing the animal with food and absorbing much of the animal's waste. We discuss one such symbiotic relationship in Issue 16.

It is no accident that to a certain extent waste from protists and plants (e.g., oxygen gas) is essential for animals and that waste from animals (e.g., nitrogen and phosphorus) is essential to plants and protistans.

Marine plants include the familiar vascular plants like those on land, having specialized root and circulatory systems for transporting nutrients, and light-sensitive chloroplasts in specialized structures called leaves to produce energy and food. Seagrasses and mangrove trees are examples; however, there are other types of marine plants that are perhaps even more important: the green, red, and brown algae. Algae absorb nutrients directly from seawater so they have no need for specialized root systems.

A list of important marine animals will include many you have probably never heard of. Here are a few. Copepods are members of the Phylum Arthropoda, a group that includes insects on land. Copepods have been called "insects of the sea." They are almost unimaginably numerous in many planktonic communities and are key herbivores in many environments, like the North Atlantic. Another important arthropod is the group known informally as krill. They are shrimp-like organisms about a centimeter long and present in vast schools in the Southern Ocean with many fascinating adaptations for life in surface water marine environments, especially the frigid-water environments of the Antarctic. Krill are featured in Issue 14. Figure 7-3 shows important copepods.

We describe important organisms more fully as needed in the text.

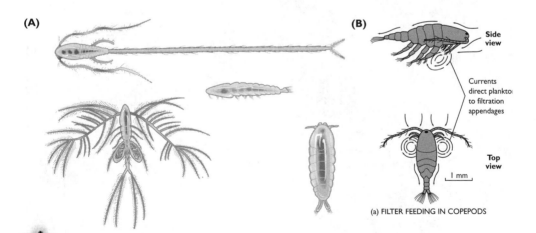

FIGURE 7-3 (A) Examples of planktonic Copepods. [*Source:* Adapted from Sverdrup, H.U., et al., *Their Oceans: Their Physics, Chemistry, and Biology.* Prentice-Hall, 1970.] (B) How copepods feed. Describe the adaptations copepods have evolved to avoid sinking.

Productivity, Nutrients, and Biomass

Regions in the oceans where abundant plant and protist life are found are said to have high primary productivity, which is a measure of the rate of production of organic matter by plants from inorganic matter. It is typically expressed in mass of Carbon per unit area per year, as "150 g C/m²/yr." Because plant growth depends on nutrients, as well as sunlight, regions of high primary productivity tend to have periodically high, or slow and steady, nutrient supply. This usually means shallow water habitats along coasts. Regions with few primary producers are said to have low productivity. Herbivores feed on the plants and thereby pass energy to larger carnivores when they are eaten. Here's one example: In the North Atlantic Ocean, one species of copepod, *Calanus finmarchicus* (Figure 7-4), typically makes up three quarters of the zooplankton biomass (the mass of organisms in a particular area at a particular time) from New England to Northern Europe. Their concentrations can reach 100,000 within each square meter of the temperate North Atlantic!

Some environments with few apparent plants still may have high productivity because protists often live inside the tissues of some animals in a relationship called symbiosis. One example is the colonial corals that build coral reefs, as we briefly mentioned previously here. Corals are tiny animals with tentacles surrounding the mouth for

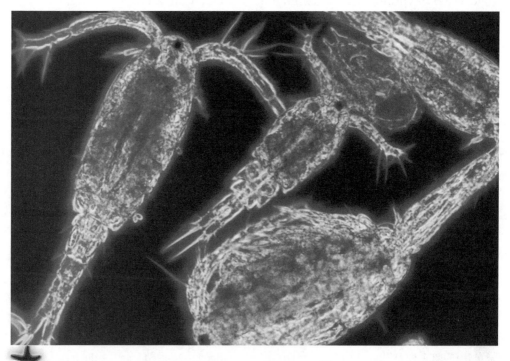

⭐ FIGURE 7-4 The copepod *Calanus finmarchicus*. This is one of the most common creatures in the ocean.

capturing and engulfing prey. Those that build rocky reefs have microscopic protists (called zooxanthellae) living within their tissues providing the corals with nutrients that supplement their diet. Scientists have recently learned that as water temperatures increase beyond 30°C, the corals may expel the protists, threatening the long-term viability of the entire reef. Thus, as the shallow ocean warms, corals and the extremely productive community of which they are the center may be increasingly at risk. We return to this subject at the end of the chapter.

Food Chains, Food Webs, and Marine Productivity

Relationships between primary producers, primary consumers (herbivores), and carnivores are shown by means of food chains and more complex food webs. At the base are primary producers, eaten in turn by herbivores and so forth. By these means is energy from photosynthesis and chemosynthesis spread throughout the myriad creatures of the marine environment.

Marine food chains and food webs label what eats what in an ecosystem. Figure 7-5 shows a food chain and a more complex food web to the consumer organism.

QUESTION 7-1. What food is eaten directly by adult herring?

It is important to emphasize that transfers of energy from one level in a food chain to another are not very efficient. For example, up to 90% of the energy in phytoplankton eaten by herbivores can be dissipated in respiration, digestion, activity, and heat. Thus, the more levels there are in a particular food chain, the less energy will be available to the largest consumers (highest level) in the chain.

Food webs can look complicated, but do not be put off: They contain very useful information. For example, Figure 7-6 shows a simplified representation of the Arctic food web.

QUESTION 7-2. How many different organisms feed on ringed seals?

QUESTION 7-3. What are the animals that feed on benthic invertebrates?

QUESTION 7-4. What is the food of the beluga whale?

QUESTION 7-5. What is the importance of the Capelin sand eel to this food web?

In this example, there were several large animals that fed on only one prey. A decline in prey numbers could have dire consequences for the predator. On the other hand, animals that feed on numerous prey species could more easily survive a drop in the abundance of one of them, although it is not known what the long-term ecological consequences would be.

As discussed in Issue 14 the Antarctic food web is extremely efficient because there are few intermediate steps between primary producers (phytoplankton), krill (herbivores), and large consumers (filter-feeding, or "baleen" whales).

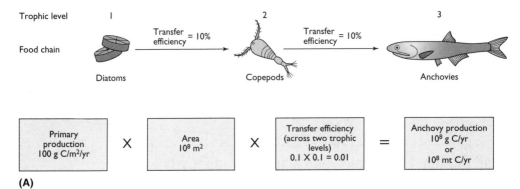

(A)

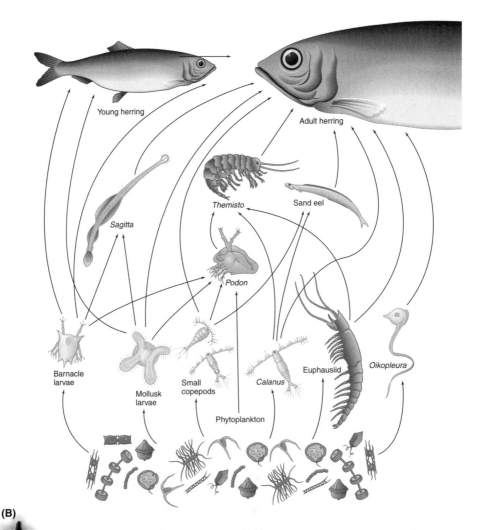

(B)

FIGURE 7-5 (A) A Food Chain. What are the members of this simple food chain? (B) A Food Web. What are the two foods of young herring? What is the difference between a food chain and food web? [*Source:* Adapted from A.C. Hardy, *Fisheries Investigations* 7 (1924): 1-53.]

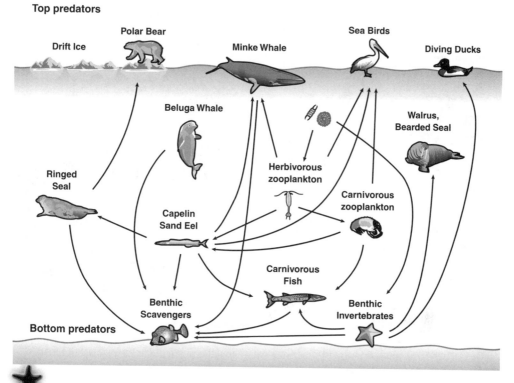

Top predators

Drift Ice

Polar Bear

Minke Whale

Sea Birds

Diving Ducks

Beluga Whale

Walrus, Bearded Seal

Ringed Seal

Herbivorous zooplankton

Capelin Sand Eel

Carnivorous zooplankton

Carnivorous Fish

Benthic Scavengers

Benthic Invertebrates

Bottom predators

FIGURE 7-6 The Arctic Food Web. What is the principal food of the polar bear? What is the principal food of the bearded seal?

Global Patterns of Marine Productivity

Figure 7-7 shows marine primary productivity. The highest primary productivity is in zones of upwelling and around continents where nutrient supply is highest. Also, note the zones in the Antarctic with very high productivity. Finally, lowest primary productivity is in portions of the open ocean, but because it is the largest area by far, the open ocean is ironically responsible for most ocean primary productivity. The oceans make up 70% of the Earth's surface, the continents the remainder. High productivity zones are upwelling zones, salt marshes, coral reefs, and mangroves, the latter three at extreme risk from pollution, climate change, overfishing and/or development.

Adaptive Strategies of Marine Organisms

An adaptive strategy, or adaptation, can be anything that helps an organism survive in a particular environment and has a genetic basis—that is, it can be passed on to offspring. Anything that helps an individual species member survive will probably help the species as a whole. Adaptive strategies may be reproductive, behavioral, morphological (shape), physiological, and/or biochemical. Here are examples.

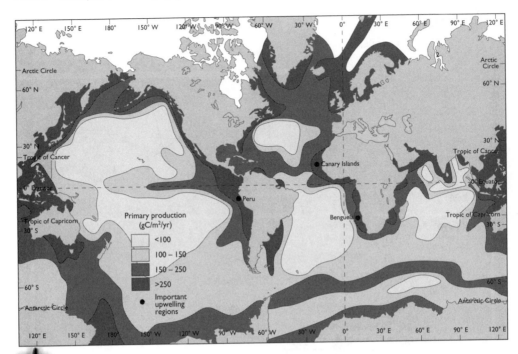

FIGURE 7-7 Marine Primary Productivity. Where in general are the areas of lowest primary productivity? Propose a hypothesis to explain why this is so. [*Source:* Adapted from Couper, A., ed. *The Times Atlas of the Oceans.* Van Nostrand Reinhold, 1983.]

■ Example 1: Reproductive Adaptations of Larvae

Many marine species have a larval stage that can be readily dispersed by ocean currents. For example, most benthic invertebrates that affix to surfaces as adults have a free-floating larval stage that spends part of its life in the plankton. The barnacle is an example.

Barnacles are notorious to seafarers for their ability to attach themselves strongly to solid surfaces, like the bottom of a boat or a dock piling (Figure 7-8). In the past they were periodically scraped off the boat by the fisher, usually timed to clean the boat before the animals had a chance to calcify and attach strongly, but the use of toxic chemicals, such as tributyl tin in antifouling paint to poison animals that attach to boat hulls has meant that such chemicals are now routinely detected in seawater, especially in ports. The long-term impact of this is unknown, but cannot be positive. The larvae of barnacles go through two stages that average a few days each and then search for a suitable attachment site. During the larval stages, they may be carried hundreds of km by currents.

Barnacles have both male and female reproductive organs (i.e., they are hermaphrodites, a condition that is not uncommon among many groups of animals) but usually must cross-fertilize. After attachment, they can begin reproducing in a few weeks and can produce thousands of eggs several times a year. The eggs, like the larvae, are dispersed by currents, although most of the eggs are eaten by predators before they develop into larvae. Most larvae are also eaten before they can develop into adults.

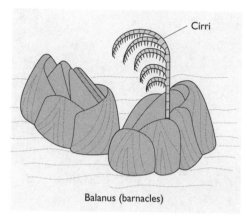

FIGURE 7-8 Barnacles. A toxic chemical, tributyl tin, has often been used to remove barnacles from ships and prevent their attachment. [*Source:* Adapted from Tait, R.V. and R.S. DeSanto. *Elements of Marine Ecology.* Springer-Verlag, 1972.]

■ Example 2: Morphological Adaptations of Diatoms

Most, perhaps 90%, of all photosynthetic organisms in the marine environment are protists. One of the most widespread types is the diatom. Diatoms are single-celled protists with a shell made of "silica," or hydrated SiO_2, called a frustule (can you see now why Si is a nutrient in seawater?). Some diatoms are circular in plan (meaning seen from the top), and some are pinnate, or elliptical. Figure 7-9A shows a sketch of a diatom, and Figure 7-9B shows a centric diatom (size 80 micrometers, or 0.003 inch). Figure 7-9C shows a pinnate diatom (about 100 micrometers long).

Planktonic protists must avoid sinking to survive. Diatoms remain in the photic zone through a series of adaptations, most notably (1) storing oil droplets, which are less dense than seawater, (2) having body shapes with spines, and (3) forming long chains. They are kept in the surface zone by the turbulent mixing of the shallow surface waters. Diatoms reproduce by cell division, and the cells can divide once every 12 to 24 hours. One limiting factor on plant growth is the availability of nutrients. In areas where nutrient influx is high, diatoms numbers can quickly reach astronomical proportions. When this happens, the phenomenon is called a diatom "bloom." In Arctic regions, such as the Alaskan coast, during the summer, the sun shines nearly all day. Meltwater streams bring nutrients into coastal waters. These are ideal conditions for diatoms, and concentrations can briefly reach the astounding figure of 12,000,000 diatoms per milliliter of seawater.

QUESTION 7-6. How many diatoms would be found in one liter (recall that there are 1,000, or 10^3 ml, in a L)?

QUESTION 7-7. How many diatoms would be found in each cubic meter of seawater?

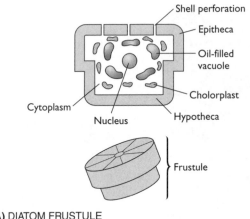

(A) DIATOM FRUSTULE

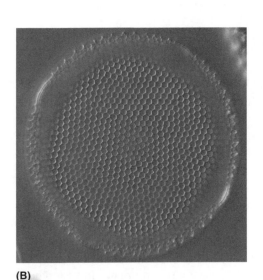

(B)

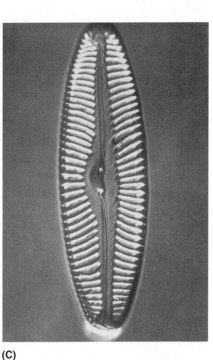

(C)

FIGURE 7-9 Diatoms. (A) This diatom secretes a frustule (shell) of silica. What are three essential marine nutrients? (B) A centric diatom and (C) a pennate diatom. Which of these is most likely planktonic? Why?

QUESTION 7-8. Finally, assume a diatom bloom covered 1 km² to a depth of one m. How many diatoms would make up this bloom (recall that 1 km² = 1,000 m × 1,000 m = 1 million m²)?

Diatoms are so common in the oceans (and elsewhere) that some scientists believe that up to one quarter of all the carbon in living tissue is in diatoms.

■ Examples of Biochemical and Behavioral Strategies

Some marine organisms are slow moving and long-lived and have relatively few offspring, but they may be very successful. One reason could be that they are toxic to potential predators or simply an impossible mouthful. The lionfish and the porcupine fish are two examples. The porcupine fish (*Diadon hystrix*) is a slow-moving nocturnal predator. In the larval stage (up to 20 cm long) it is free-swimming, and it is very vulnerable to predators; however, in the adult stage, it is vulnerable only to large predators such as sharks. This is because it has rows of sharp spines on its body that lie flat under normal conditions. If the fish is disturbed, it can take in water and instantly assume the shape of a ball of spikes. All but the largest predators and the slowest learners are dissuaded from eating it. Figure 7-10A shows a porcupine fish.

Figure 7-10B shows a lionfish, which are venomous coral reef fishes that have recently been observed along the East Coast of the United States, probably having been introduced to the Florida Coast first, likely dumped from aquaria. Although no deaths from lionfish have been reported, they can cause extremely painful stings.

"Dispersal of the lionfish population along the Atlantic coast was likely helped by Gulf Stream transport of lionfish eggs and larvae," said NOAA scientist Jonathan Hare, who co-authored an assessment report on Atlantic lionfish.

QUESTION 7-9. Look again at Figure 7-10B. Describe the appearance of the lionfish. Is it well-camouflaged or distinctive?

A venomous organism that could mistakenly be eaten by a predator often evolves adaptive strategies that make it conspicuous so predators will recognize it and avoid it. Moreover, there is no need for venomous fishes to be able to swim rapidly to avoid predators. You can see this in the body appendages of the lionfish.

■ Convergent Evolution in Lamnid Sharks and Tunas

A particularly striking example of adaptation occurs in two groups of fishes, the lamnid sharks and the tunas (Figure 7-11). These organisms also exhibit a phenomenon known as convergent evolution, which describes the possession of similar adaptations by organisms that are not closely related. Although both groups are fishes, the sharks are cartilaginous (i.e., without a bony skeleton) fishes, whereas tunas are bony fishes groups that diverged over 400 million years ago. The shark family Lamnidae (pronounced "**Lam**-nid-ee") contains the great white, mako, porbeagle, and salmon sharks, whereas the bony fish family Scombridae (**Scom**-brid-ee) includes the bluefin, skipjack, yellowfin, and albacore tunas. Consider the mako shark and albacore tuna, both among the oceans' top predators.

Superficially, the mako shark and albacore tuna resemble each other (Figure 7-11). Both species are robust, highly streamlined, heavily muscled fish. Their tails are lunate (crescent shaped), a design that provides powerful thrust. Their similarities go deeper as well. Both species are endothermic, meaning that they can trap some of the heat produced during metabolism and elevate their core body temperature above that of their

(A)

(B)

FIGURE 7-10 (A) A porcupine fish; (B) A lionfish. These fish are often toxic. How do they "advertise" their toxicity?

environment. This is not an easy task for aquatic creatures, the overwhelming majority of which are ectotherms, meaning that their body temperature is the same as that of the environment. Water very efficiently removes heat from anywhere it contacts a body warmer than it is. Marine mammals can retain some heat by having thick blubber or fur as insulation. They also breathe air; thus, little heat is lost during respiration. Diving

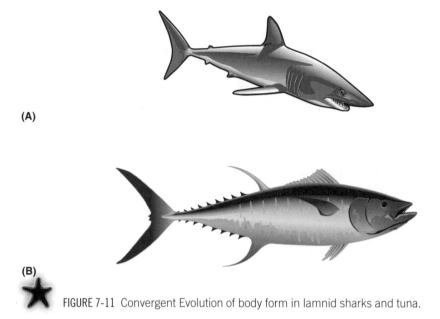

(A)

(B)

FIGURE 7-11 Convergent Evolution of body form in lamnid sharks and tuna.

birds have oily feathers and/or fat to prevent heat loss, but fish tend to lose heat through their body and especially their gills. So how do the mako shark and albacore conserve heat? They use a simple engineering design known as countercurrent exchange. Blood vessels coming from the hotter body core where excess heat is produced when muscles contract containing heated blood run to the body's periphery surrounded by blood vessels containing colder blood coming in the other direction, from the periphery. The heat is very efficiently transferred across the thin walls of the blood vessels; thus, as the colder blood proceeds inward to the body core, it is heated, and the hotter blood has lost its heat to the incoming colder blood before it could lose it to the seawater. The anatomical structure of these blood vessels is known as a rete mirabile (pronounced **ree**-tee mī-**rab**-ī-lee), meaning wonderful network.

What is the advantage of a warmer body? A warmer body means that chemical reactions occur more quickly, which results in more powerful muscle contractions and thus faster swimming speeds. Additionally, sensory information is processed faster. The combination of better senses and faster swimming speed makes these organisms among the oceans' best predators. Moreover, both species are able to move into colder water without experiencing the problems (e.g., less muscle power) that most sharks or bony fishes would. Endothermy, however, comes with a cost. The mako shark and the albacore tuna must eat a large amount of food to power their metabolic machinery.

■ Summary: Adaptive Strategies

There are perhaps as many different kinds of adaptive strategies in the ocean as there are different kinds of organisms. Some examples of strategies we have had no time to describe in detail are (1) fish that develop large 'eye spots' which confuse predators;

(2) fish that look like, attach to, and behave like strands of algae; (3) organisms like octopi that can change the color and patterns of their skin and thus blend in with their background; (4) fish that can mimic the color and even texture of the sandy bottom; (5) krill, fish, and squid that school in vast numbers which minimizes the chance that any one individual will be eaten and facilitates reproduction; (6) crabs that crop living algae and attach it to their shells and serves as camouflage; (7) fish that develop body shapes and skin textures which minimize drag; (8) colonial corals that spawn simultaneously at times keyed to maximum tides, which maximize fertilization, dispersal, and survival of eggs; (9) copepods living at great depths with very large, light-sensitive eyes; and (10) deep-water fishes with bioluminescent "lures" attached to their head, which entice prey.

Those who have easy access to the Internet can find a wealth of fascinating information beyond our brief introduction.

Climate Change and Marine Ecology

Over millennia, organisms evolve adaptations that help them survive in their environment, at least to an age that allows them to reproduce and propagate the species. To understand the potential impact of climate change on organisms in the oceans, you need to appreciate (1) how organisms become adapted to particular environments and (2) how and at what pace climate change will alter these environments.

One theory of evolution holds that organisms adapt to their environment gradually over many thousands of generations. If change in the environment takes place slowly, that is, at a pace that would allow evolutionary modifications to compensate for the change, many organisms might do very well. If on the other hand, as most scientists predict, the rates of change will be greater than rates at which evolution can produce modifications to compensate for change, then the future for many marine species will be dire indeed. Dire, that is, if we allow a "business as usual" approach to addressing the drivers of climate change.

Here are the kinds of changes that an episode of global warming will impose on the oceans.

- The surface water will warm rapidly, and the polar seas will warm most rapidly.

- Polar seas may experience a rapid decline in salinity from melting sea ice, glaciers, and ice caps, which would have profound effects on ocean circulation. An increase in surface fresh water from melting ice could slow, or even stop, the clockwise circulation of warm, tropical water from the Caribbean to the North Atlantic. We discuss this in Issue 6.

- Sea level will rise from higher water temperature as well as from melting ice caps and glaciers (but not from melting sea ice). We consider this in Issue 5.

- Coastal ecosystems, like highly productive salt marshes and mangroves, will be flooded by gradually rising seas, at rates that will likely exceed their ability to compensate by migrating inland. This, coupled with development pressure (from construction and aquaculture), could wipe out many of these ecosystems. Stresses to coastal ecosystems are featured in Issues 1–3, 10, 11, 16, and 17.

- Increases in atmospheric CO_2 will lead to increases in CO_2 dissolved in seawater. Other things being equal, this could potentially enhance marine productivity; however, other things are never equal. Adding CO_2 to seawater would first lead to acidification of shallow seas, and later on (hundreds of years later) to acidification of deeper waters. Acidification will make it harder or impossible for many $CaCO_3$-secreting organisms to produce shell or skeletal matter. It could further hasten the demise of many coral reef communities, already at grave risk from pollution and development. These are among the most productive and diverse ecosystems on Earth. Threats to coral reefs are considered in Issue 16.

- Solubility of gases in seawater is inversely related to temperature. As shallow sea temperature rises, oxygen solubility could decline. Impacts of eutrophication from hypoxia (low oxygen) and anoxia (no oxygen) are considered in Issue 11.

- Increase in surface ocean temperature increases the energy available to power tropical disturbances like hurricanes. These storms can cause more coastal destruction in one event than during decades of normal processes.

Finally, the history of the Earth has been a history of change. The issue is not whether marine organisms can adapt to change or whether climate change will occur. The issue is rather whether unprecedented rates of change brought about by human interference in natural cycles (Milankovic, continental fragmentation and assembly, sea floor spreading rates) will overwhelm the ability of organisms to respond to these changes. We ask that you keep this fact in mind as you work your way though this book.

PART 3

Human Population Growth

Population Growth: Overview and Global Trends

How does human population growth threaten the marine environment?

What is the Earth's population and how fast is it growing?

Where is population growing fastest?

What is population projected to be in the future?

What can we do to lessen these threats?

Who is responsible for solving the problem?

Point-Counterpoint: Is Population Growth a Problem?

Globally, the population debate is about whether Earth has exceeded its human carrying capacity, the maximum number of people that the planet's resources can sustain in the long-term without irreversibly harming the planetary ecosystem. What is Earth's carrying capacity, and have we exceeded it?

One view is that the carrying capacity of the planet is essentially limitless. Economist Julian Simon, whose views are supported by many mainstream economists and others, argued that the Earth could support 20 billion or more people. A chapter in his book *The Ultimate Resource II: People, Materials, and Environment* is entitled "Can the Supply of Natural Resources—Especially Energy—Really Be Infinite? Yes!"

He wrote this:

Adding more people causes problems, but people are also the means to solve these problems. The main fuel to speed the world's progress is our stock of knowledge, and the brakes are (a) our lack of imagination, and (b) unsound social regulation of these activities.

The ultimate resource is people—especially skilled, spirited, and hopeful young people endowed with liberty—who will exert their wills and imaginations for their own benefit, and so inevitably benefit not only themselves, but the rest of us as well.

Simon's view is disputed by many scientists and environmentalists. They cite evidence that suggests we have already exceeded the planet's carrying capacity. For example, renowned biologist Garrett Hardin wrote this:

On the American scale of living the carrying capacity of the Earth is only about one-hundredth as great as it would be if people would be content with the barest minimum of goods....

Moreover, some [amenities] impose costs that cannot be stated... the solitude of lonely beaches, access to wilderness and areas rich in flowers, birds, and butterflies, together with time to enjoy these amenities as well as music and the visual arts....

Thus does the problem of the optimum human population become inextricably woven with problems of value.

Pioneering oceanographer Jacques Cousteau, inventor of SCUBA diving, was more pessimistic:

The road to the future leads us smack into the wall: ... a demographic explosion that triggers social chaos and spreads death, nuclear delirium and the quasi-annihilation of the species... Our survival is no more than a question of 25, 50 or perhaps 100 years.

After you have read the following chapter, and indeed the entire book, judge for yourself. Which view is more nearly correct, or are there are alternative scenarios? If human population growth is a serious problem, what can be done to ameliorate it?

Introduction

Population growth is central to the entire field of environmental studies. Some, however, might question whether human population growth is a marine environmental issue in the strictest sense. Here are our reasons why human population growth is included as the first issue in this environmental oceanography text. Most of these are covered in greater detail in the issues that follow.

Although humans exert a substantial positive impact with our enormous ingenuity, we also have a profound physical impact on the marine environment, and the marine environment also impacts us.

First, we take up space—space that at one time was natural, such as maritime forest, salt marsh, mangrove swamp, or hillside at the edge of a bay. The space an individual occupies may be minimal, as in cities in New Guinea (Figure I 1-1A), or it may be large, as in newer U.S. single family houses, which averaged over 230 m^2 (2,500 ft^2) by 2008, or even larger, as in beach houses (Figure I 1-1B).

Nutrient pollution, one of the most pervasive forms of water pollution, kills fish in streams and estuaries (Figure I 1-2), causes large-scale dead zones in the oceans, and is largely responsible for the environmental demise of Chesapeake Bay. These nutrients come from fertilizers used on agricultural crops and lawns, pets in urban areas, animal waste from so-called factory farms (also known as animal factories or concentrated

FIGURE I 1-1(A) Human population growth encroaching on the coast at Port Moresby, New Guinea.

FIGURE I 1-1(B) A typical southeastern US beach house, typically erected on or slightly behind the primary sand dune line, the beach's first line of defense against a storm. [Photo by D. C. Abel.]

FIGURE I 1-2 Neuse River, North Carolina fish kill after Hurricane Floyd. How or why did these fish die?

animal feeding operations) that raise livestock for human consumption, and treated (and frequently untreated) human waste.

People also require transportation, and in the United States, that overwhelmingly means roads. Roads generate polluted runoff that can include heavy metals and hydrocarbons. Roads also take an enormous toll on animals (including marine creatures ranging from fiddler crabs to endangered sea turtle hatchlings), divide habitats (which can accelerate species loss), and provide human access to sensitive places like barrier islands and other wild places, frequently allowing the invasion of noxious plant species (e.g., by seeds affixing to tires).

The waste created by the growing number of humans is another major environmental problem. Globally, human-generated sewage and commercial, residential, and industrial wastes increase in volume each year. In the United States, each person produces nearly 763 kg (1,700 pounds) of municipal waste per year. This does not count sewage, mining, and other industrial waste. Some of this waste, which includes toxic substances and pharmaceuticals that have passed through humans, inevitably enters the marine environment.

Availability of water for domestic, agricultural, and industrial uses is also affected by population growth. Both surface and groundwater sources are becoming endangered by pollution and excessive withdrawals. In some areas, removal of freshwater causes localized intrusion of saltwater into aquifers used as water sources for human consumption. Large-scale desalination plants built to convert salt water into clean drinking water threaten coastal ecosystems with their waste.

Freer (less regulated) world trade and a more integrated world economy mean citizens of one country can have an increasing impact on the environment in another. Large-scale, and often illegal, fishing by ships from Europe, China, and South Korea is replacing small-scale, less destructive local fisheries along the coast of West Africa (see Issue 18). The result is a collapse of the local economies, a loss of subsistence fishing, and disruption of the marine food web that could lead to crashes of fish populations. It is also driving illegal migration of fishers and their families to Europe.

Moreover, demand for raw materials such as tropical hardwoods or ores by developed countries can lead to accelerated erosion and transport of sediment and toxic materials into marine ecosystems.

Nonindigenous species (also called exotic or alien species) have successfully established themselves in estuaries such as San Francisco Bay and Chesapeake Bay, having hitchhiked across oceans in the ballast water of cargo ships and tankers (see Issue 22). In their new location, these introduced species can harm ecosystems by disrupting food webs and can cause extensive economic damage—for example, by impacting fisheries or physically clogging industrial water intakes.

Humans also require energy, and for the next two decades at least, that means fossil fuels, which impose a cost on the marine environment when they are extracted, refined, transported, and burned. Much of this cost is not reflected in the price we pay for energy or products, representing a substantial subsidy that further encourages the wasteful use of fossil fuels. The prospects for energy from offshore wind, waves, and tides appear good, but these come with potential enviromental impacts as well (see Issue 24).

This ever-increasing human population is concentrating in coastal areas at risk from hazards like sea level rise and tropical storms. At least 60% of the planet's human population lives within 100 km of the coast, and coastal cities are reaching sizes unprecedented in history (e.g., Sao Paulo, Brazil's population is approaching 20 million; see Issue 2). Fifty-two percent of the U.S. population lives in coastal counties. Thus, ever-increasing numbers of people are exposed to flooding from tropical storms, as well as to coastal erosion and sea-level rise accompanying global climate change (see Issue 5).

The growing popularity of ocean cruises, especially to threatened and sensitive marine areas such as coral reefs, Alaskan bays, and the Antarctic places additional stress on the oceans from fuel spills, discharged human wastes, illegally dumped shipboard waste, and disturbance and destruction of habitat. For example, many marine vessels currently use the most polluting, highest-sulfur form of oil, called bunker fuel, or fuel oil 6. These ships are already a major source of marine air pollution (see Issue 9).

Growing wealth in developing countries means greater demand for animal protein, increasingly from species that may already be overfished (see Issues 12 and 18).

As you can see, human population growth is the quintessential marine environmental issue.

. . . .

Exponential Growth and Its Impacts

On October 12, 1999, the United Nations "celebrated" its "Day of 6 Billion," that is, the day at which the Earth reached an estimated human population of 6 billion. According to the U.S. Census Bureau, the Earth's population was 5 billion in 1987, 4 billion in 1974, 3 billion in 1959, and 1 billion in about 1825. Before we can analyze the effect of the impact of population growth, it is necessary to understand the simple math that describes such growth.

When any quantity such as oil consumption, the human population, or a bank account grows at a fixed rate (or percentage) per year, say 1% or 10%, that growth is said to be exponential (Figure I 1-3). If a quantity grows by a fixed amount every year, say 500,000 barrels, or 80 million people, or $1,000, the growth is said to be arithmetic. Exponential growth can be described as growth upon growth and, given enough time, will be explosive, whereas arithmetic growth is stepwise and more gradual.

■ Calculating Exponential Growth the Easy Way

Calculating exponential growth can be intimidating to those of us with math anxiety or who are rusty in the use of math. Fortunately, there are ways to make growth calculations that involve only addition, subtraction, multiplication, and/or division. These can provide relatively accurate estimates but have some limitations.

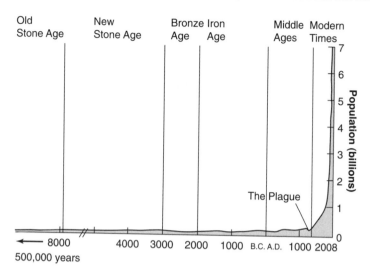

FIGURE I 1-3 World population growth, which exhibits a J-shape, or exponential, curve. Is it possible for human population growth to continue forever? Why or why not?

Precise calculations of exponential growth are made using the compound interest formula (also known as the compound growth equation), the same one used to calculate interest on funds in bank accounts. The compound growth equation allows population scientists (demographers) to project future population size, assuming a current population and population growth rate. Using it is not as intimidating as it may seem at first and is demonstrated in Appendix 1.

Exponential growth, including population growth, is based on this simple mathematical relationship:

Growth rate over a period = (population at end of period − population at beginning of period)/population at beginning of period

Replacing words with symbols, we get this:

$$r = (N - N_0)/N_0$$

The variable r = growth rate is expressed as a decimal (e.g., 5.2% would be entered as 0.052); N = the population at the end of the period; and N_0 (pronounced "N sub-zero") = the population at the beginning of the period.

This works if you want the growth rate for an entire period, say 1 year or 50 years; however, as we show you in Appendix 1, it is less accurate than the compound growth equation in some cases (e.g., if you want to know the annual growth rate over a multiyear period, for example 50 years).

Let us first estimate (demographers would use the term *project*, like we did earlier) the human population for the future. Here is an example:

The mid-year 2006 world population was 6.52 billion and was growing at a rate of 1.14%. Project what the world population would be in 2020.

To answer this, we use:

$$N = (r \times N_0 \times \text{the number of years}) + N_0$$
(which is a modification of the formula presented above)

Recall that 1.14% = 0.0114, and 6.52 billion = 6.52×10^9 (consult Appendix 1 if you need a quick refresher on using scientific notation).

Substituting numbers for symbols:

$$N = [0.0114 \times (6.52 \times 10^9) \times 14] + (6.52 \times 10^9)$$
$$N = 7.56 \times 10^9, \text{ or } 7.56 \text{ billion}$$

If you had used the more accurate compound growth equation, the more precise answer would have been 7.65 billion. For our purposes, the first answer is close enough, although you should be aware that the size of the error increases with longer periods of time and higher growth rates.

QUESTION 1-1. The mid-year 2008 world population was 6.8 billion and was growing at a rate of 1.17%. Project what the world population would be in 2025 and 2050 at this constant growth rate. Use either $N = (r \times N_0 \times \text{the number of years}) + N_0$, or the compound growth equation from Appendix 1.

Both the simpler growth equation and the compound growth equation can be rearranged if you want to know the average growth rate or how long it would take a population of a given size to grow (or decrease) to a different size at a specified growth rate. Here is an example of the former (once again, consult Appendix 1 for the rearrangement of the compound growth equation if you wish to use it):

Given a 1987 world population of 5 billion and a 1999 world population of 6.0 billion, calculate the average annual growth rate over the 12 year period.

To answer this, we use this:

$$r = [(N - N_0)/N_0]/\text{\# years}$$

Substituting numbers for symbols:

$$r = [(6 \times 10^9) - (5 \times 10^9)/(5 \times 10^9)]/12$$
$$= 0.0167, \text{ or } 1.67\%$$

Using the compound growth equation, which you will recall is more accurate, your answer would have been 0.0152, or 1.52%. Again, for our purposes, this gives us a reasonable approximation.

QUESTION 1-2. Given a 1987 world population of 5 billion and a 2008 world population of 6.8 billion, calculate the average annual growth rate over the

21 year period. Use either $r = [(N-N_0)/N_0]/\text{\# years}$, or the compound growth equation (Appendix 1).

Here is an example of how long it would take a population of a given size to grow (or decrease) to a different size at a specified growth rate:

At an annual growth rate of 1.14%, how long would it take a population of 5 billion to grow to 6 billion?

This time we use:

$$\text{\# years} = [(N-N_0)/N_0]/r$$

Substituting numbers for symbols:

$$\text{\# years} = [(6 \times 10^9)-(5 \times 10^9)/(5 \times 10^9)]/0.0114$$
$$= 17.5 \text{ years}$$

The answer obtained from using the compound growth equation is 15.99 years. We also have a useful, quick, easy, and relatively accurate rate to do similar calculations, known as doubling time.

■ Doubling Time

For any population that is growing exponentially, the time it takes for the population to double is calculated using the equation $t = 70/r$. In this doubling time formula, in contrast to the other growth formulas, r is entered as the decimal growth rate times 100; for example, a growth rate of 0.07 (i.e., 7%) is entered as 7. The doubling time formula is explained and derived in Appendix 1.

QUESTION 1-3. Use the doubling time formula to estimate how long it would take a population of 6.8 billion to double in size given an annual growth rate of 1.17%.

QUESTION 1-4. Does a population growth rate of 1.17% sound large? Explain then why it might be cause for concern. Be as specific as you can and cite examples if possible.

Understanding the impact of population growth is more than just a numbers game. Biologist Garrett Hardin's "Third Law of Human Ecology" states that the total impact of a population on its environment is determined by the absolute population size multiplied by the impact per person.

QUESTION 1-5. In our calculations, we have thus far ignored the impact per person. In what ways does your impact on the planet as a U.S. resident differ from your perception of the impact of a resident of a developing country? Think of your diet, and focus especially on impacts on the marine environment.

QUESTION 1-6. In light of your answer to Question 1-5, what do you think would be most effective in minimizing the impacts of humans on the planet: controlling population growth in developing countries (where most growth is in fact occurring) or curbing consumption of resources in developed countries (where the impact per person is highest)? Cite evidence for your answer. Identify the assumptions you used to answer this question.

QUESTION 1-7. Upholding and improving women's rights and interests (including educational, reproductive, economic, and healthcare) have been called the most important actions to control population growth. Explain how this might work. Do you agree or disagree with the statement? Give evidence for your answer.

QUESTION 1-8. Summarize the main points in this Issue.

Most projections of human population growth suggest that the population will reach 9 billion by 2050. After that, growth rate is projected to decrease, but the population could continue to grow, although at a slower pace.

We started this issue by bringing up the idea of human carrying capacity. As you complete the issues in this book, continuously ask yourself if environmental problems caused by humans are evidence that we have exceeded our human carrying capacity. If so, then our future quality of life, or perhaps even survival, depends on sharply reducing our impacts and/or decreasing our population.

■ Media Analysis

Go to www.npr.org, and search for "634 Million People at Risk from Rising Seas."[1] Listen to the report; then read the accompanying text on the web page, and answer the following questions.

The implication of the researchers is that 634 million people are at risk from rising seas, but they have not considered the impact of population growth. Let us do such an analysis for the 10 countries in the two groups the researcher listed.

QUESTION 1-9. List the ten countries with the most people in the low coastal areas.

QUESTION 1-10. List the countries with the largest share of their population living in low elevation areas.

QUESTION 1-11. Which countries are on both lists?

QUESTION 1-12. What steps do you think these countries could take to lessen the impact of rising sea level?

QUESTION 1-13. Go to https://www.cia.gov, and search for "CIA World Factbook."[2] Determine the present population and population growth rates for the three countries on both lists.[3] Fill in the appropriate spaces in Table I 1-1.

---

1. If you search for the report using the title, make sure that you have selected "Sort: Relevance" and not "Sort: Date." The choice will appear in the middle-right of the web page. Or go directly to http://www.npr.org/templates/story/story.php?storyId=9162438.
2. Or go directly to https://www.cia.gov/library/publications/the-world-factbook/.
3. Select "Guide to Country Profiles." Pick the country from the pull-down menu, and then click on "People" on the right side of the screen.

TABLE I 1-1	Population data for countries with the most people in the low coastal areas and with the largest share of their population living in low elevation areas (from www.cia.gov)				
Country	Approximate Population (mid-2007)	Growth Rate (%)	Doubling Time (70/r)	Population after one doubling time	Population by 2100

QUESTION 1-14. Calculate the doubling times for each country's population based on the published growth rates, and place them in Table I 1-1. Doubling time is calculated using $t = 70/r$.

QUESTION 1-15. Use the doubling time equation to roughly estimate their populations by the year 2100, at which time sea level is projected to have risen 60 cm (2 ft). [Alternatively, use the compound growth equation (Appendix 1) to project their populations by 2100.]. Fill in Table I 1-1.

QUESTION 1-16. How do these calculations change your assessment of the threat of sea level rise to these countries?

Coastal Population Growth: Bangladesh and Miami, FL

KEY QUESTIONS

How does human population growth threaten the coastal ocean?

How can we measure these threats?

What can we do about them?

Who is responsible for addressing the problem?

To what extent, if any, is coastal population growth compatible with a sustainable ocean?

Analysis 1: Bangladesh

Bangladesh (Figure I 2-1), a country about the size of Illinois, Wisconsin, or Florida, lies on the northern shore of the Indian Ocean.

Bangladesh was originally part of Pakistan after the Indian subcontinent gained independence from Britain in 1947. In 1970, Bangladesh seceded from Pakistan. Its population in mid 2007 was nearly 151 million on a total area of 144,000 km^2 (55,000 mi^2). The mid 2007 population of the United States was about 301 million on a total area of 9.8 million km^2 (377,000 mi^2).

QUESTION 2-1. What is the population density of Bangladesh in m^2/person (hint: there are 1,000^2, or 10^6, m^2 in a km^2)?

QUESTION 2-2. What is the population density of the continental United States in m^2/person?

■ Floods in Bangladesh

Bangladesh is vulnerable to catastrophic flooding from rivers as well as from tropical storms. More than half the country lies at less than 8 meters (26 feet) above sea level. Moreover, more than 17 million people live on land that is less than 1 meter (3.3 feet)

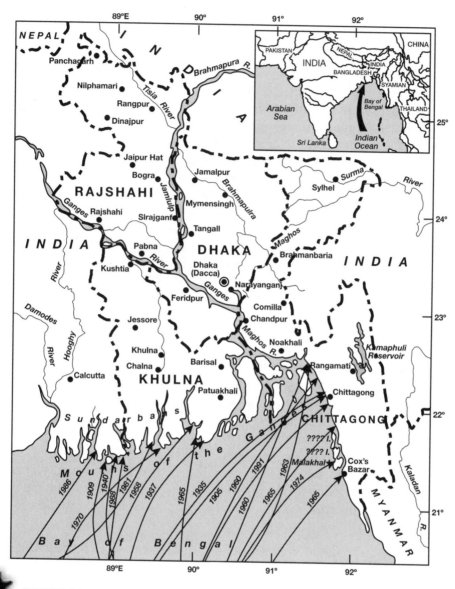

FIGURE I 2-1 Location of Bangladesh, showing tracks of major historical typhoons.

above sea level. As a result, about 25% to 30% of the country is flooded each year. This can increase to 60% to 80% during major floods.

River flooding in Bangladesh results from some combination of the following: (1) excess rainfall and snow melt in the watershed, especially in the foothills of the Himalayas, an area where many of the Earth's rainfall records were set; (2) simultaneous peak flooding in all three main rivers; and (3) high tides in the

Bay of Bengal, which can dam runoff from rivers and cause the water to "pond" and overtop its banks.

In addition, changes in land use in the catchment area can result in profound changes in the flood potential of Bangladesh's rivers. For example, deforestation of the Himalayan foothills for agriculture, fuel, or habitation makes the hills more prone to flash flooding. Then, during the torrential monsoon rains, mudslides can dump huge volumes of sediment into the rivers. This sediment can literally pile up in the river channels, reducing channel depth and thereby the channel's capacity to transport water. The result is more water sloshing over the rivers' banks during floods and more disastrous floods in Bangladesh. The silt can smother offshore reefs when it finally reaches the ocean, made worse by land-use practices that clear coastal mangrove forests. Geologists estimate that the drainage system dumps at least 635 million tonnes (1.4×10^{12} pounds) of sediment a year into the Indian Ocean, which is four times the amount of the present Mississippi River.

Western-style development may be contributing to Bangladesh's increased vulnerability to natural disasters. Recently built roads blocked floodwaters, thereby increasing flooding from cyclones and the damage and loss of life.

Truly catastrophic floods can result when high tides coincide with tropical storms. In the early 1990s, such a combination took over 125,000 lives. About one tenth of all Earth's tropical cyclones occur in the Bay of Bengal (Figure I 2-1), and approximately 40% of the global deaths from storm surges occur in Bangladesh.

Storm surges result when cyclones move onshore and the winds push a massive wall of seawater onshore with them. Surges in excess of 5 meters (16.4 feet) above high tide are not uncommon. You can imagine the impact such a storm can have on a country in which half the land is within 8 meters (26.2 ft) of sea level!

QUESTION 2-3. The population of Dhaka, Bangladesh was 8 million in 1998. First, calculate the doubling time (i.e., when the population will increase to 16 million) at the 1998 growth rate of 6% per year, using the doubling time formula, t = 70/r (see Appendix 1). The population was estimated at 10.8 million in 2008. How did this agree with the value you calculated from your use of 1998's growth rate?

QUESTION 2-4. Using the 2008 growth rate of 3.9%, estimate when the population of Dhaka will reach 21.6 million (i.e., double the 2008 population).

QUESTION 2-5. Bangladesh is a trivial producer of CO_2 compared with nations like the United States and China, and it faces potential disaster from sea level rise because of global warming and tropical storms. What responsibility, if any, do you believe the main carbon-emitting nations have to reduce the potential for environmental disaster in Bangladesh?

QUESTION 2-6. Discuss whether you think protecting marine ecosystems should be a priority for Bangladesh. What measures could they take?

QUESTION 2-7. Assume that the United States is at least partially responsible for the potential environmental disaster in Bangladesh. List as many actions as you can, by priority (if you can), that the United States could take to prevent, postpone, lessen, or clean up environmental damage and resultant destruction of property and loss of life.

QUESTION 2-8. Summarize the main points of this analysis.

. . . .

Analysis 2: Miami FL
. . . .

Miami, Florida is literally and figuratively a world away from Bangladesh. It is the play-ground of the super wealthy, a popular vacation spot, as well as the home of large num-bers of expatriates, mainly Cubans, from Caribbean countries. The fates of both places, however, are inexorably tied to global climate change. Table I 2-1 shows population data for Miami from 1960 to 2007.

QUESTION 2-9. Determine annual rate of population growth for Miami for the period 2000–2007. Use the equation $r = [(N - N_0)/N_0]/\#$ years, introduced on page 94–95.

QUESTION 2-10. Use the annual growth you calculated in Question 2-9 to determine the doubling time for Miami's population. Use $t = 70/r$ (see Appendix 1).

QUESTION 2-11. Using the doubling time you just calculated to project what the population will be in three doubling periods, starting in 2000.

QUESTION 2-12. Miami/Dade County had a 2006 population of about 2.4 million on a land area of 6,000 km^2. Calculate the city's population density in m^2 per person (recall that $1 \, km^2 = 10^6 \, m^2$). Compare it with that of Bangladesh from Question 2-1.

QUESTION 2-13. Examine Figure I 2-2A and I 2-2B. Given the likelihood that a 1.5-m sea level rise could occur as early as the turn of the century and that a 5-m increase in sea level is projected to follow if the Greenland Ice Sheet melts, what measures would you take now to prepare?

QUESTION 2-14. Summarize the main points in this Issue.

TABLE I 2-1   Population for Miami, Fl from 1990–2007

2007: Miami/Dade County, 2,387,170
2000: Miami/Dade County, 2,253,362
1990: Miami/Dade County, 1,937,094

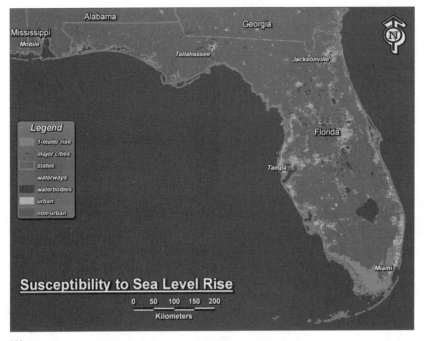

(A)

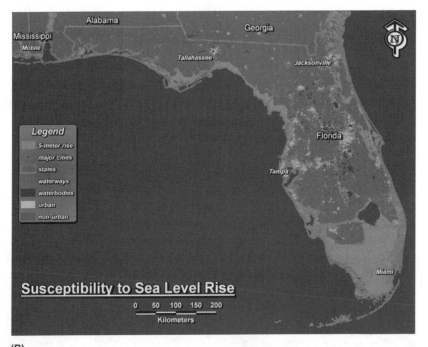

(B)

FIGURE I 2-2 A/B (A) Representation of a 1.5 M sea-level rise in Florida; (B) Representation of a 5 M rise.

Media Analysis 1

Go to www.npr.org and search for "Biscayne Bay" (remember to use the quotation marks).[1] Scroll down to "Miami Plans Home for Mega-Yachts," and listen to the 4-minute program. Then answer the following questions.

QUESTION 2-15. Where is the proposed development to be situated?

QUESTION 2-16. What "need" is the development planned to satisfy?

QUESTION 2-17. How did former Miami official Carlos Jimenez describe his view of the proposed site without the development? How do you respond to his characterization?

QUESTION 2-18. Recall Chapter 5 on sustainable oceans. What type fuel do marine vessels use? Assess the impact on the local Biscayne Bay environment after the development has opened for business.

Media Analysis 2

Go to www.npr.org and search for "Florida Bay" (remember to use the quotation marks).[2] Scroll down to "Massive Florida Algae Bloom Leads to Accusations." Accelerating population growth in South Florida as well as intensive agriculture with its accompanying polluted runoff is putting at risk one of the last remaining pristine nearshore environments in Eastern North America. Listen to the 5-minute program, and answer the following questions.

QUESTION 2-19. What is the problem facing Florida Bay described in the program?

QUESTION 2-20. What part of the Bay is affected?

QUESTION 2-21. Describe the bloom and its effect on bottom seagrasses.

QUESTION 2-22. How is the local Bay's food chain affected by the bloom?

QUESTION 2-23. How could the Florida Transportation Department have contributed to the problem?

QUESTION 2-24. What has been the response of the Transportation Department to these allegations?

QUESTION 2-25. What was the observed response of the bloom to the reduction of sunlight in the winter?

QUESTION 2-26. Describe the lingering effect of the 1989 algal bloom on another part of the Bay.

QUESTION 2-27. What has been the practical effect six years after passage of the "massive" Everglades Restoration Plan?

---

1. Or go directly to http://www.npr.org/templates/story/story.php?storyId=6355921.
2. Or go directly to http://www.npr.org/templates/story/story.php?storyId=6707247.

What's Up, Dock? Unsustainable Coastal Development

How many people live along the coast?

What are some of the problems associated with coastal growth?

How has coastal development influenced the beach environment?

What, if anything, can or should be done about coastal erosion?

What are the pros and cons of beach-front stabilization?

What is the impact of beach renourishment on fauna?

Point-Counterpoint

On the subject of the artificial maintenance of beaches, Chris Brooks, South Carolina's former top coastal regulator said: "You've got intense development on the shore, and you've got to keep the beach in front of them.... I'd like to say there are alternatives to beach renourishment, but from a practical standpoint, it's about the only answer we've got."

Coastal geologist Orin Pilkey demurred: "Nobody is looking at the big picture. Sea level is rising and we expect storms to increase. Nourished beaches will get more difficult and costly to maintain. Long term, it's pointless anyway. Nature is going to win at the shoreline."

What information do you need to analyze and evaluate the issue?

Introduction

Chances are you live within 100 km (62 mi) of the coast. According to the World Resources Institute, at least 60% of the planet's human population lives that close to the coastline, and coastal areas have the fastest growing populations. Coastal cities are reaching sizes unprecedented in human history. Not surprisingly, over half of the world's coastlines are at significant risk from development related to this population growth.

Fourteen of the 20 most populous cities in the United States are on the coast, and all of Australia's major cities—Sydney, Melbourne, Brisbane, Adelaide, and Perth—are coastal. Coastal counties account for only 17% of the land area of the contiguous United States; however, 16 million more people live in these counties than in the 83% of inland counties. Moreover, this population is increasing by at least 3,600 per day. U.S. coastal population is projected to increase to 166 million in 2015.

Look at Figure I 3-1, which shows projected changes in U.S. population between 1994 and 2015.

QUESTION 3-1. Where is growth projected to be most rapid?

QUESTION 3-2. Go to Issue 2 and review Figure I 2-2 (impact of sea level rise on Miami and Florida). Comment on the long-term implications of growth in these areas on the marine environment.

QUESTION 3-3. Do you think insurance companies will continue to provide insurance to homeowners in light of sea level increases? If so, what do you project will happen to rates? If not, how will this impact the responsibilities of the state and federal governments?

QUESTION 3-4. Discuss whether you think that state and/or federal government can afford to guarantee rebuilding costs for coastal dwellers.

QUESTION 3-5. How did government's response to Hurricane Katrina affect your answer to Question 3-4?

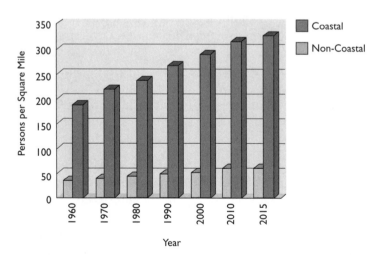

FIGURE I 3-1 Population growth of coastal and non-coastal constituents of the U.S. from 1960 to estimated values in 2015. Data from NOAA State of the Coast Report, 1998. [*Source:* Data from NOAA State of the Coast Report, 1998.]

. . . .

Interfering with Coastal Processes

. . . .

We now return to the issue raised in "Point Counterpoint:" whether it makes sense to protect shorelines against the effects of wave attack.

Recall from Chapter 6 that beaches are "sand rivers," in which the sediment is transported by longshore currents. Moreover, erosion of beaches may be intense during infrequent storms. Beaches are dynamic, frequently unstable zones that change in shape over both space and time and thus may not always be the best sites for development; however, the demand for beach-front homes is at an all-time high.

Over time, the shape of beaches is influenced by short-term storm events (days), normal seasonal changes in wave intensity (months), longer term climatic phenomenon like El Niño and La Niña and periods of hurricane activity (years to decades), and very long-term (century to millennium) changes in global sea level.

In the last 18,000 years, the sea level has risen about 120 meters (400 feet), and many coastal shorelines have moved far inland. Off the East Coast of the United States geologists estimate that the shoreline has migrated inland more than 80 km (50 mi) during this period, alternating between active migration and stasis. For each unit rise in sea level, the beach will migrate 1,000 to 2,000 units inland.

QUESTION 3-6. Study Figure I 3-2, which shows a projected 60-year erosion hazard map, for South Bethany, Delaware. Using the scale, determine how far inland the beach is projected to erode over the next 60 years. What is the average rate expected in feet per year? How many houses will be affected?

QUESTION 3-7. If sea level is projected to rise by about 60 cm by 2100, how far inland on average will the coastline move?

If you add to this phenomenon the anticipated increased hurricane activity in the Atlantic, then the long-term prognosis for coastal development begins to look precarious.

QUESTION 3-8. A conservative estimate of the projected rise in sea level by 2100 is 60 cm (2 feet). How do you think the typical American perceives the seriousness of a sea level increase of this magnitude over a 90-year period? How do you think this affects their conclusion about the urgency of the problem?

Coastal erosion is a global problem. Along the Nile Delta of Egypt, shoreline erosion averages 50 to 100 meters per year and occasionally rises to 200 m/year, largely because of the trapping of Nile River sediment behind the Aswan Dam.

In the United States, all 30 states bordering an ocean or the Great Lakes are experiencing coastal erosion, and 26 are presently experiencing net loss of their shores. According to the Federal Emergency Management Agency, approximately 1,500 homes will be lost to erosion annually. Losses of this magnitude or greater are projected to continue for the next several decades. The Federal Emergency Management Agency estimates the current cost of this loss to be at least $530 million each year.

Inland limit of
60-year erosion hazard area

20 0 20 40 60 Feet

Site Location

N

 FIGURE 13-2 Aerial photo and projected erosion map of South Bethany, Delaware.

Composition and Structure of Beaches

A grain of sand, by definition, is between 0.06 and 2.0 mm in diameter. There are an estimated 7.5×10^{18} (7.5 quintillion) grains of sand on all of the Earth's beaches. Where any one of those grains is going to be at any time is a question of immense significance to coastal dwellers, tourists, policy makers, and even insurers.

Sand nourishes beaches. It can come from a variety of sources, including adjacent beaches (whose dunes represent a reservoir of sand), rivers, eroding cliffs, and the continental shelf. Geologists consider all of the sources of sand for a particular area as a single, linked system rather than as isolated, independent supplies.

The shape of a beach varies with sand supply, sea level change, and wave size. When any of these factors change, the other factors respond accordingly. For example, a reduced supply of sand, with sea level and wave size remaining constant, will cause the landward movement of the beach's intertidal region. Decreasing wave energy with the other physical factors constant will cause either reduced erosion or natural accretion (building) of the beach.

Rates of Coastal Erosion

The actual rate of shoreline erosion depends on supply of sand, wave energy, sea-level change, and nature of coastal bedrock.

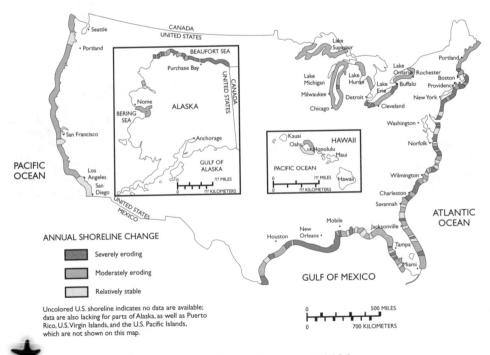

FIGURE I 3-3 Coastal erosion rates. [*Source:* Courtesy of USGS.]

Average annual erosion rates are 2 to 3 feet/year on the Atlantic coast and about 6 feet on the Gulf coast (Figure I 3-3). During intense storms, 50 to 100 feet of Atlantic and Gulf coast shore can erode.

QUESTION 3-9. In your judgment, where in Figure I 3-3 is the erosion problem most severe? We are not asking where erosion is most severe but where the problems it causes are most severe. Under what assumption did you carry out this activity?

QUESTION 3-10. In the light of limited government revenue and the desire of many to relentlessly reduce taxes and the role of government, to what extent is it an essential responsibility of government to pay, out of general revenue, to protect coastal infrastructure? If it is a responsibility of government, at what level should this responsibility lie? Explain your reasoning, or cite evidence.

A Brief History of Coastal Protection Measures in the United States

The Federal Rivers and Harbors Act, passed in 1889, gave the U.S. Army Corps of Engineers the responsibility for maintaining navigable waterways, including harbor entrances. The Corps' mandate was gradually extended to protecting coasts. It responded by building seawalls and groins (discussed later here). These structures temporarily protected coasts but resulted in loss of beaches in front of the structures. Coastal dwellers then demanded beach renourishment, that is restoration of the lost beaches.

■ Coastal Protection Structures and Strategies

Humans build structures along coasts for two reasons: to trap sand and to protect the shore and what we have built on it from erosion; however, human attempts to slow beach erosion often exacerbate the problem. Construction of seawalls, jetties, revetments, groins, and other structures (Figure I 3-4) may stabilize local portions of the beach but can shift those problems to adjoining areas. Having said this, engineering structures and practices can protect coasts from erosion. The following strategies or structures have been used.

Groins and Jetties
Groins (Figure I 3-5) and jetties (Figure I 3-6) are structures, usually concrete or rocks, built at angles to coasts to trap sand. Jetties are similar structures, built to protect harbors. Beaches are dynamic, changing features, and the sand comprising them is almost always in motion from waves and longshore currents. Groins and jetties are built to keep sand in place, to protect property values, or to protect the economic value of coastal vacation destinations.

Breakwaters
Breakwaters are structures built parallel to shores to absorb wave energy, to protect the shore behind the breakwater, and/or to provide a safe harbor for ships. Figure I 3-7 gives an example.

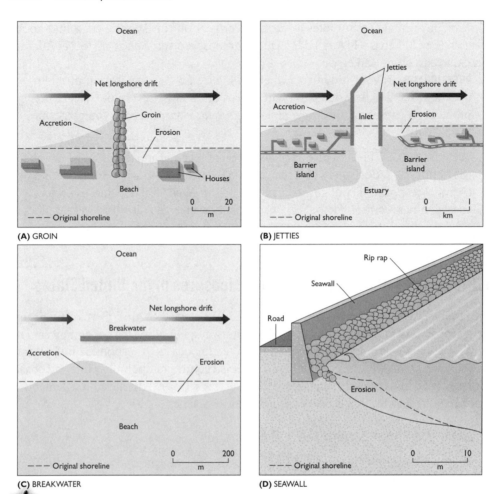

FIGURE I 3-4 Engineered structures used to stabilize shorelines. (A) Groins (B) Jetties (C) Breakwater (D) Seawall.

Sea Walls and Revetments

Seawalls (Figure I 3-8) are usually built when nothing else will keep the area behind the wall from eroding, but the beach then disappears. They can also increase erosion at adjacent beaches. The inevitable loss of the beach is known as "Newjerseyization," named for the state where seawall construction has been a common practice in postponing property loss.

Revetments (Figure I 3-9) are structures on coastal bluffs that absorb the energy of incoming waves. When properly designed, they do not significantly interfere with longshore transport, nor do they redirect wave energy to adjacent unprotected areas. Table I 3-1 summarizes the advantages and disadvantages of coastal engineering.

QUESTION 3-11. Who benefits from coastal engineering projects?

 FIGURE I 3-5 People sitting on a groin at low tide at Pawleys Island, South Carolina. [Photo by D. C. Abel.]

FIGURE I 3-6 Jetties. Explain how these structures "work."

Beach Replenishment

Although many states no longer permit construction of hardened structures such as sea walls and groins, the alternative practice of beach replenishment, or "renourishment," has become an increasingly common in the last several decades. Beach renourishment involves bringing in sand by dredge, barge, or truck to build up the height and breadth of disappearing beaches. Beach renourishment is practiced at numerous vacation destinations along the eastern United States. During winter months, high-energy waves and storms carry sand away from beaches and move it farther offshore. Sand thus removed is no longer available to nourish beaches. The sand for replenishment has to come from somewhere. Removal of sand from some site for dumping elsewhere usually destroys animal and plant communities living in the sand. The sand may come from inland sources or from subtidal shelf regions.

This practice provides a key benefit compared with constructing hard structures in that it allows the beach to continue to exist. This benefit of beach renourishment, combined with a significant federal subsidy, is an especially appealing alternative in areas that depend on tourism. The U.S. Army Corp of Engineers (COE) is responsible for designing and implementing renourishment operations. Much controversy exists regarding the COE's renourishment practices. One of the most common problems is that the COE overestimates how long a beach will last after each renourishment. The COE estimates the cost of renourishment to vary from $350,000 to $3 million per mile along the Atlantic Coast.

 FIGURE I 3-7 Breakwater. Explain how this structure "works."

FIGURE I 3-8 A sea wall stabilizing a developed shoreline along a South Carolina estuary. The area behind the seawall is protected from eroding, but the beach disappears. [Photo by D. C. Abel.]

FIGURE I 3-9 A revetment, left. Also, note the seawall to the right. [Photo by D. C. Abel.]

Renourishment practices have led to a false sense of security for many beach-front residents, and the continued subsidy of renourishment by the federal government has had the effect of stimulating additional beach-front development. Technically, we may be able to continue in the near future to repair the localized storm impact erosion to prime tourism locales (win the battle), but the reality of global sea level rise means that in the future we will eventually have to abandon current beach-front property (losing the war) or install an extremely expensive series of dikes to protect our shoreline. The Dutch spent around $15 billion during the 1990s, which is around $1,000 for each citizen, suggesting that with sea-level rise we might have to spend more than $300 billion to "protect" our coasts!

TABLE I 3-1   **Advantages and disadvantages of coastal engineering practices and strategies**

- **Advantages**
  They can protect coasts over short periods of time from erosion.
  Structures are generally built and maintained by agencies of government whose budgets are not dependent on local taxation.

- **Disadvantages**
  They are ineffective over time periods of several decades to centuries.
  They often result in sand loss.
  They encourage a sense of complacency and security, and encourage additional investment and population growth.

QUESTION 3-12. Consider the Point-Counterpoint at the beginning of this Issue. First, explain both Brook's and Pilkey's positions.

QUESTION 3-13. Discuss which of these positions you support. Support your conclusion with evidence.

QUESTION 3-14. Summarize the main points in this issue.

Other Strategies to Protect Coasts

Besides hard stabilization methods and renourishment, other strategies include zoning, relocation of structures, and legislation of various types. Zoning ordinances may restrict building in certain areas. They may require certain types of building or may require structures be set back an appropriate distance from the shore. In extreme cases, structures may be relocated, and areas may be protected from development by zoning or by legislation.

. . . .

Media Analysis 1

Go to www.npr.org and search for "coastal erosion."[1] Scroll down to a 2007 story, "La. Wants to Change River's Course to Save Coast," and listen to the 5-minute program. Then answer the following questions.

QUESTION 3-15. How rapidly is the Louisiana coast losing land?

QUESTION 3-16. How has human engineering of the Mississippi changed the river's pattern of sediment deposition?

QUESTION 3-17. The state is creating land by dredging silt from shipping channels and dumping it into wetlands. What is the ratio of land "saved" to land that is lost by erosion?

QUESTION 3-18. How much sediment is being dumped into the deep Gulf every year?

QUESTION 3-19. Describe the proposal to change the course of the Mississippi to protect the coast.

QUESTION 3-20. Where are the funds to change the River's course proposed to come from?

QUESTION 3-21. What is the expected impact of the project if carried out on the state's oyster industry?

---

1. Or go directly to http://www.npr.org/templates/story/story.php?storyId=10785780.

Media Analysis 2

Go to www.npr.org and search for "beach renourishment."[2] Scroll down to the 2007 story, "Importing Sand, Glass May Help Restore Beaches," and listen to the 8 minute program. Then answer the following questions.

QUESTION 3-22. How many times has sand been placed on the John U. Lloyd State Park's beaches since 1970?

QUESTION 3-23. What was the source of the sand used to renourish the beach?

QUESTION 3-24. How much sand was used?

QUESTION 3-25. Why can't the beaches recover on their own?

QUESTION 3-26. Discuss whether you agree with allowing a new condo on the sand that was trapped by the jetty and that thus never reached the state park's beach.

QUESTION 3-27. Why is beach restoration not just a cosmetic procedure?

QUESTION 3-28. Discuss whether you agree or disagree with Charlie Finkle's statement, "We need to keep renourishing the beach to keep up with the rise in sea level and if we don't keep up with the renourishment, we're going to fall behind and then you're going to have bigger disasters."

QUESTION 3-29. List and evaluate the options engineers are considering.

QUESTION 3-30. Was "doing nothing" mentioned as an option? What are the advantages and disadvantages of doing nothing?

QUESTION 3-31. Coastal geologist Orrin Pilkey has said, "Nature bats last at the shoreline." What do you think he means?

---

2. Or go directly to http://www.npr.org/templates/story/story.php?storyId=12026379.

PART 4

Global Climate Change

What Causes Climate Change?

What causes climate change?

What gases are mainly responsible?

What is the relationship between CO_2 and temperature based on Antarctic ice cores?

How is the Milankovitch Theory related to climate change?

What are effects of climate change on the oceans?

KEY QUESTIONS

Introduction

A scientific consensus has been reached that increasing levels of atmospheric "greenhouse gases," mainly CO_2 and methane are driving global temperatures to levels not seen on this planet for more than 600,000 years. The gases are mainly from extracting, transporting and burning the fossil fuels coal, oil, and natural gas, but also are from the clearing of forests (deforestation) that occurred first in China and Europe, later in North America, and most recently in the tropics and high latitudes. More importantly, these changes are occurring at rates that dwarf climate change rates observed over the past 600,000 to one million years.

Misunderstandings about the role of CO_2 and other "drivers" of climate change have led some skeptics to dismiss the role of CO_2. They point out that, based on ice core records in the Antarctic, the onset of higher atmospheric CO_2 lagged behind the onset of higher temperatures by 200 to 800 years (Figure I 4-1). Skeptics therefore suggest that some other mechanism, and not CO_2, caused and, more importantly, is causing global climate change.

In this issue, we investigate the climate change record over two time frames. The first is the glacial–interglacial climate change record of the past 600,000 years, and the second is the historical record, that is, the most recent 1,000 years of climate change (Figure I 4-2).

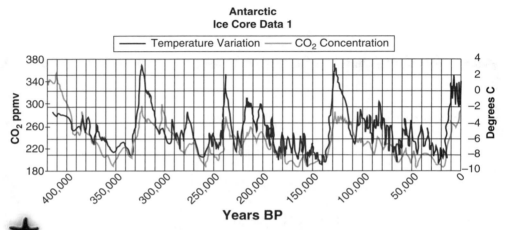

FIGURE I 4-1 Data from Antarctic ice cores, showing the correlation between temperature and CO_2 in bubbles of gas within the ice.

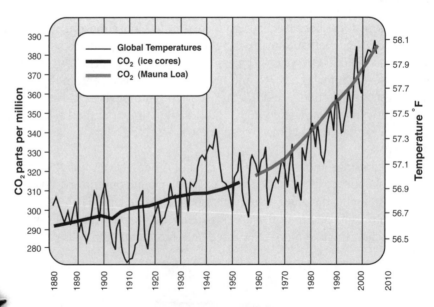

FIGURE I 4-2 Temperature and CO_2 variation since 1880.

QUESTION 4-1. Describe the association between CO_2 and temperature for the past 120+ years.

To understand the controversy, we need to introduce and explain one of the most important and least appreciated of all scientific theories: the Milankovitch Theory.

Milankovitch Cycles and Climate Change

Milutin Milankovitch was a successful civil engineer before he was offered a professorship in mathematics in Belgrade, Serbia in 1909. He then began work on the variability of the Earth's orbit and its potential to control the Earth's climate. By 1941, he had concluded that variations in the Earth's orbit were the primary causes for the formation and retreat of ice sheets over the past three million years, and perhaps over the past 2.5 billion years. Geological evidence suggests that the Earth has experienced at least four Ice Ages: one centered on about 2.5 billion years, one at around 700 million years, one around 280 million years, and one beginning around 2.5 million years before the present. Refer back to Figure I 4-1. Most geologists believe that, before the past 200 years changed everything, we were in an interglacial period headed toward another glacial episode.

Using elegant mathematics and detailed observation, Milankovitch identified three cyclical variations in the Earth's orbit, each occurring at differing time scales. The variations are as follows:

1. Orbital eccentricity—that is, the change in the way the Earth orbits around the Sun. The Earth's orbit is not a perfect circle—at times, far from it. When the orbit is highly elliptical (in the shape of an oval), the amount of insolation (i.e., the rate of solar radiation per unit area) received at closest approach would be 20% to 30% greater than at farthest distance.

   QUESTION 4-2. How would the Earth's climate be different between a circular orbit and a strongly elliptical orbit?

   Currently, the Earth comes closest to the Sun around the first week of January, but the northern hemisphere is inclined (tilted) away from the Sun. Thus, when the northern hemisphere experiences winter and receives the least amount of sunlight, the Earth as a whole receives the most. The eccentricity of the Earth's orbit varies with a 100,000-year cycle (Figure I 4-3a and I 4-3b).

2. The second factor is change in obliquity, which is the change in the angle between the Earth's axis and the plane of the Earth's orbit (Figure I 4-3b). The Earth's axial tilt varies between 22.1° and 24.5° with a 41,000-year period.

3. The third factor is precession, which is the "wobble" of the Earth's axis over time (Figure I 4-4). The axis makes about one full circle in 26,000 years.

   The interaction of these three cycles determines global climate, other things being equal, according to Milankovitch. Now, usually these three cycles do not of course perfectly coincide, but at certain fateful times in Earth history, these three cycles have

(A)

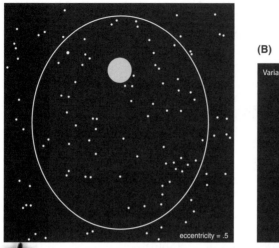

(B)

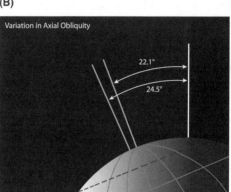

Variation in Axial Obliquity

22.1°

24.5°

eccentricity = .5

 FIGURE I 4-3 (A) The most extreme elliptical nature of the Earth's orbit around the Sun. About 100,000 years from this picture, the orbit will be nearly circular. (B) The variation in Earth's orbital obliquity. [*Source:* NASA.]

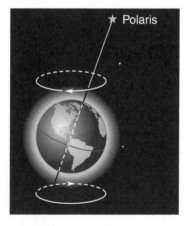

★ Polaris

FIGURE I 4-4 The precession of the Earth's axis.

coincided so as to magnify the effect that each one might have. When they coincide, an extremely warm episode or extremely cold episode is likely to occur, the latter possibly leading to the Earth's great ice ages. In addition, the onset of colder climates at the Poles is likely to generate a positive feedback in the following manner. After permanent snow and ice begin to accumulate, the white matter will increase the albedo, that is, the reflected radiation, reducing the solar energy absorbed by the surface and strengthening the cooling episode.

Milankovitch cycles have been interpreted from rhythmically banded sedimentary rocks from the far geologic past. So geologists are very confident that these cycles have been affecting Earth's climate for at least the past three hundred million years. Of course, whether Milankovitch cycles cause large-scale changes in terrestrial climate, such as glaciation, depends on many other factors, such as the size and position of continents.

Impact of Milankovitch Cycles on Global Climate: Summary

Milankovitch cycles control variation of global climate by increasing and decreasing the amount of solar insolation received by the Earth. The coincidence of all three cycles causes maxima and minima of solar insolation and, thus, other things being equal, more extreme warming and cooling episodes of terrestrial climate. These cycles operate on time frames of tens to hundreds of thousands of years and are responsible for the long-term glacial–interglacial cycles that have dominated terrestrial climate for the past several hundred thousands of years. Warming the planet causes ice sheets to melt and soils to thaw, both of which release CO_2 to the atmosphere. Additionally, CO_2 is less soluble in water as it warms. Thus, a warmer ocean also releases CO_2 to the atmosphere. Cooling the planet causes the reverse.

Next, we briefly discuss a short history of research on climate change.

Historical Research on Climate Change

■ John Tyndall

In 1857, physical scientist John Tyndall began research on what he called "the radiative properties of gases." He discovered that there were differences in the ability of atmospheric gases to absorb "radiant heat," which is what he called electromagnetic radiation in the infrared range. He found that oxygen and nitrogen, which make up almost 99% of the atmosphere, had virtually no ability to absorb infrared. Two atmospheric gases, CO_2 and water vapor, H_2O, were powerful infrared absorbers. He concluded that without these two gases the Earth's surface would be "held fast in the iron grip of frost."

He speculated on whether changes in the concentrations of CO_2 and H_2O could affect the Earth's climate.

■ Svante Arrhenius

In 1895, the distinguished Swedish chemist Svante Arrhenius, winner of a 1903 Nobel Prize in Chemistry, presented a paper to the Stockholm Physical Society titled "On the Influence of Carbonic Acid in the Air upon the Temperature of the Ground." Building on the work of Tyndall and others, he argued that variations in CO_2 and H_2O could profoundly affect the Earth's heat budget. His calculations led him to conclude that "the temperature of the Arctic regions would rise about 8 or 9 degrees Celsius if the [CO_2] increased 2.5 to 3 times its present (1895) value."

■ Roger Revelle

If there is a "father of global warming," it is marine geologist Roger Revelle (1909–1991). In a pioneering article published in 1957, Revelle and colleague Hans Seuss showed that CO_2 concentration in the atmosphere had increased as a direct result of the burning of fossil fuels and that the oceans were unlikely to be able to absorb all of the increase. They built their work on earlier studies, including the work of British engineer Guy Stewart Callendar who, using refined analytical techniques, advanced a hypothesis that warming measured between 1900 and 1938 could be the result of rising fossil fuel burning.

Until 1957, many scientists recognized the importance of the greenhouse effect but thought that humanity was unlikely to be able to alter global climate. Their reason was that the oceans store 50 times as much CO_2 as the atmosphere. If this relationship were to be constant, only 2% of any additional CO_2 released by human activities would remain in the atmosphere: The rest would have been dissolved in the oceans.

In 1957, Revelle and Seuss demonstrated that the oceans could not absorb CO_2 as rapidly as humanity was releasing it. In their paper's conclusion, they used one of the most memorable phrases in the literature of science: "Human beings are now carrying out a large-scale geophysical experiment" on the Earth.

Revelle and Seuss's hypothesis led to the establishment of monitoring stations on Mauna Loa, Hawaii and in Antarctica to measure worldwide trends in atmospheric CO_2 concentrations. In 1965, a President's Science Advisory Committee concluded that Revelle and Seuss had been correct and officially recognized global warming from CO_2 as a possible global problem. The CO_2 concentration plot from Mauna Loa has become one of the most recognizable icons in science (Figure I 4-5).

By 1977, Revelle and his colleagues had concluded that 40% of the human-generated CO_2 remained in the atmosphere. Two thirds was from fossil fuels, the rest from deforestation. They speculated that the oceans were absorbing much of the "missing" CO_2, but not nearly so much as the 50:1 ratio previously supposed.

■ Climate Change Comes Out of the Closet

On June 23, 1988, during a Washington, D.C. heat wave, Dr. Jim Hansen, head of NASA's Goddard Research Institute for Space Studies, testified before a Senate Committee. Here is part of his testimony.

I would like to draw three main conclusions. Number one: the Earth is warmer in 1988 than at any time in the history of instrumental measurements. Number two: global warming is now large enough that we can ascribe with a high degree of confidence a cause-and-effect relationship to the greenhouse effect. And number three: our computer climate simulations indicate that the greenhouse effect is already large enough to begin to affect the probability of extreme events such as summer heat waves.

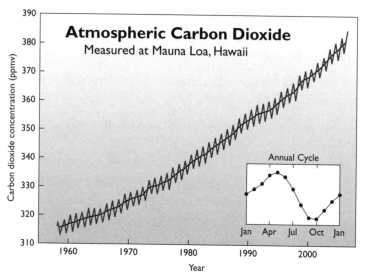

FIGURE I 4-5 Recent mean monthly variation in atmospheric CO_2 measured at the Mauna Loa observatory. "PPMV" means "parts per million by volume".

Hansen's testimony started a controversy that has only recently begun to abate with the near-universal realization that Hansen, Revelle, their colleagues, and their predecessors were right. Humans are changing the climate (Figure I 4-6).

QUESTION 4-3. Explain the dilemma posed by the "lag time" between onset of warming and increase in CO_2 during the cycles of the past 600,000 years.

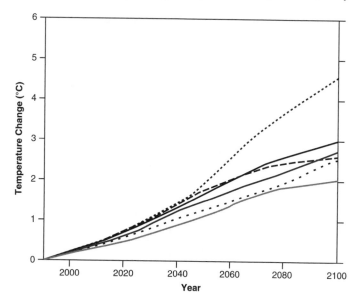

FIGURE I 4-6 Computer forecasts for climate change to 2100. What is the range in projected temperature increase for 2040?

QUESTION 4-4. Does this lag time mean that CO_2 increase from burning fossil fuels, deforestation, and so forth is not the cause of the past several hundred years of warming? Why or why not?

QUESTION 4-5. When did scientific data begin to be presented that humans were altering the climate?

QUESTION 4-6. When did Roger Revelle publish his groundbreaking papers? What was the importance of Revelle's 1977 conclusion about the ability of the oceans to absorb CO_2?

QUESTION 4-7. When did Hansen state that a "high degree of confidence" existed about the relation between climate change and human-produced CO_2?

QUESTION 4-8. What is the range of the Intergovernmental Panel on Climate Change (IPCC) projections of global temperature increase to 2100 shown in Figure I 4-6?

QUESTION 4-9. Summarize the main points in this Issue, especially the difference between the Milankovitch-induced controls on premodern climate and the human-induced climate change of the past 200 years.

Media Analysis 1

Go to www.npr.org, and search for "Antarctica Ice Melt Speeding Up." Listen to the 5-minute program, and then answer these questions.[1]

QUESTION 4-10. What is the depth of the circumpolar current (CC) described by the researcher around the edge of Antarctica?

QUESTION 4-11. What did the researcher say was happening to the CC?

QUESTION 4-12. What is the apparent effect of this change on the ice sheets on the border of Antarctica?

QUESTION 4-13. If all the ice on Antarctica were to melt, how high would sea level rise?

QUESTION 4-14. What additional data may have to be considered to revise the prevailing estimates of a 1-meter rise in sea level by 2100?

QUESTION 4-15. Explain why Dr. Rignot did not concur that the situation was "more alarming than we thought" and why he thought that the scientific community was conservative in their estimates of sea level rise. If necessary, review the introductory section on the nature of science (Chapter 1).

---

1. Or go directly to http://www.npr.org/templates/story/story.php?storyId=18112630.

- -

Media Analysis 2

Go to www.npr.org, and search for "Geologists Study Ancient Climates" (in quotation marks). Listen to the 9-minute program from 1997,[2,3] and then answer the following questions.

QUESTION 4-16. What are the objectives of the climate scientists working on the *Joides Resolution*?

QUESTION 4-17. At the location of the drill ship during the program, how different was sea level during the "Wisconsin" glacial period compared with today?

QUESTION 4-18. What was the southern boundary of the "Laurentide" ice sheet 16,000 years ago?

QUESTION 4-19. What is the raw material from which the ship is extracting information about climate and environment?

---

2. Or go directly to http://www.npr.org/templates/story/story.php?storyId=1028450.
3. Listening to this report requires RealPlayer, which is available free from www.real.com.

ISSUE 5

Sea Level Rise

KEY QUESTIONS

What factors cause changes in sea level?

How can scientists estimate the magnitude of sea level rise using the thermal expansion of seawater?

What effect will melting of ice caused by sustained global warming have on sea level?

How can we assess the severity of sea level rise's impact on coastal ecosystems and their human populations?

Introduction: An Unlikely Scenario?

At 2:06 a.m., 02/22/2025, you wake out of a deep sleep to police sirens and hear something being said on a bullhorn about evacuation. A minute later, air raid sirens join in the frenzy. You turn on a light and then the television. There are maps and satellite images of Antarctica and shots of vast ice cliffs crashing into the ocean. Incredulous, you listen to the live news report, "At 11:53 p.m. yesterday evening—slightly more than two hours ago—a large portion of the West Antarctic Ice Sheet began to slide violently into the adjoining polar ocean." You stare at the television screen, speechless. The news anchor continues, "All residents of coastal states living within 10 miles of the ocean or coastal bays are requested to evacuate immediately. The National Oceanographic and Atmospheric Administration has predicted that the water displacement generated by the slide of this ice sheet into the ocean will send a wave of 10 meters along the edge of the Atlantic Ocean basin. It is predicted that portions of coastal Florida, Georgia, and South Carolina will be impacted by this wave in roughly 48 to 72 hours. This evacuation will be permanent because the global sea level after the wave passes will be 6 meters above current sea level—we now transfer you to the White House for an emergency address by the President."

You tell yourself this must be a dream. This is impossible! But is it?

What Causes Sea Level Changes?

Water covers 71% of the Earth's surface. Geologists believe that the mass of H_2O (in all of its states) on the Earth's surface has been roughly the same for several billion years. This relative constancy of water mass does not, however, mean that sea level has remained constant. Changes in sea level have occurred repeatedly during the Earth's geologic history. Global sea-level change is caused by two processes:

Eustatic processes change the amount of liquid water within the ocean basins. The main mechanism driving eustatic change in sea level is the re-proportioning of H_2O between water and ice, caused by changes in global climate.

Isostatic processes change the underlying topography of the sea floor and can occur on either regional or global scales. Examples of regional changes include (1) the rebound of Earth's crust after melting of continental ice sheets, (2) the slow subsidence (sinking) of deltas at passive continental margins, and (3) changes in sea floor spreading rates (Chapter 6). As spreading rates at the mid-ocean ridge speed up, heat emitted at divergent boundaries increases. This causes the mid-ocean ridge to expand and push sea level higher. When rates slow, the reverse happens. Change in sea floor spreading rates cause very slow sea level changes—over millions to hundreds of millions of years.

Changes in sea level may have contributed to mass extinction events within the Earth's geologic past, and rapid sea level decline has been hypothesized to cause decline in atmospheric oxygen levels as newly exposed organic-rich marine deposits oxidize, removing oxygen from the atmosphere.

Sea level has been rising since the end of the last glacial episode about 15,000 years ago, an example of the reproportioning process we mentioned previously here. The rate of global sea level rise for the last 100 years has been about 2 mm/year. Global sea level rise for the next century could double to at least 4 mm/year and 40 to 45 cm total (0.16 in/year, 16 to 18 in total). The observed rate of coastal sea level change varies from region to region because of the variability in isostatic crustal movement (see page 44–45). If sea level rise speeds up, shallow water coastal marine communities may be unable to keep pace with sea level change and may die off as environmental conditions change beyond their limits of tolerance. Such a change is now occurring in intertidal salt marsh environments around the Mississippi River delta in Louisiana (Figure I 5-1). There, coastal subsidence caused mainly by the compaction of delta sediments combined with global sea level rise exceeds the rate of vertical growth of the marsh community.

What Are the Impacts of Global Climate Change on Sea Level?

Although scientists have documented impacts of changing sea level on biological communities from the geological past, humans have only recently become concerned with the potential of such processes to change their own lifestyles. This newfound concern

FIGURE I 5-1 Subsidence in intertidal salt marsh environments around the Mississippi River delta in Louisiana. How is this subsidence related to deltaic sedimentation?

stems from the consensus among climatologists that the planet is experiencing a period of accelerated warming caused mainly by human activity.

Global warming may result in an increase in the rate of sea level rise resulting from (1) the thermal expansion of seawater and (2) the melting of ice in glaciers and at the poles. The coefficient of thermal expansion of seawater is 0.00019 per degree Celsius, meaning that if a volume of seawater occupied 1 cubic meter of water, after warming by 1°C, it would expand to 1.00019 m^3.

Whereas thermal expansion acts on water already in the ocean, melted ice represents water added to the present ocean volume. The melting of all of the ice that is currently perched on terrestrial land, as in the ice sheets of Greenland, Iceland, and the Antarctic, has the potential to raise sea level by about 80 meters (260 feet). Ice that is already floating in the ocean water, as in the Arctic ice mass, Antarctic ice shelves, and icebergs, may melt but will not contribute to sea level rise. The mass of water contained in these features already displaces approximately its equivalent water volume.

There is abundant evidence that melt rate has increased. Monitoring of mountain glaciers over the last 2 decades shows that at least three quarters of them are losing mass. The rate of this loss has almost doubled in the last 20 years alone. In 1991, NASA reported that the extent of sea ice in the Arctic Ocean declined by 2% between 1978 and 1987. More recently, Norwegian scientist Ola Johannessen, of the Nansen Environmental and Remote Sensing Center, has used satellite measurements to show declines in permanent Arctic ice of 7% over each of the past 2 decades.

In the Antarctic, three ice shelves, Wordie, Larsen a, and Larsen b, have collapsed over the last decade, spurred by the 2.5°C increase in the average local temperature since the mid 1940s. There is some concern that loss of these ice shelves may destabilize the vast continental ice sheets behind them, with the Western Antarctic Ice Sheet potentially capable of slipping into the surrounding ocean.

How Much Will Sea-Level Rise Cost?

A recent Environmental Protection Agency report concluded that "the total cost for a one meter rise would be $270-475 billion (in the U.S.), ignoring future [coastal] development."

QUESTION 5-1. Identify an assumption in the previous quotation.

The report continues:

We estimate that if no measures are taken to hold back the sea, a one meter rise in sea level would inundate 14,000 square miles, with wet and dry land each accounting for about half the loss. The 1,500 square kilometers (600–700 square miles) of densely developed coastal lowlands could be protected for approximately one to two thousand dollars per year for a typical coastal lot. Given high coastal property values, holding back the sea would probably be cost-effective.

Which Areas Will Be Affected?

Sea level rise will increase flood risks in areas already at or under sea level, like New Orleans, Louisiana in the United States and coastal Holland in Europe. In Bangladesh, about 17 million people live less than 1 meter (3.05 feet) above sea level. In Southeast Asia, a number of large cities, including Bangkok, Bombay, Calcutta, Dhaka, and Manila, are located on coastal lowlands or on river deltas. Particularly sensitive areas in the United States include the states of Florida and Louisiana, coastal cities, and inland cities bordering estuaries.[1]

Moreover, low-lying islands in the Pacific (Marshall, Kiribati, Tuvalu, Tonga, Line, Micronesia, Cook), Atlantic (Antigua, Nevis), and Indian Oceans (Maldives) will be greatly impacted. For example, in the Maldives, most of the land is within 1 meter (3.05 ft) of sea level. A seawall recently built to surround the 450-acre capital atoll of Malè (Figure I 5-2) cost the equivalent of 20 years of the Maldivian gross national product, according to U.N. reports.

QUESTION 5-2. Given the mean depth of the ocean, 3,800 m, how much would a rise of seawater temperature of 1°C cause sea level to rise? Hint: convert 3,800 m to centimeters, and then multiply this number by the coefficient of thermal expansion?

---

1. The web site http://www.architecture2030.org/ has satellite images of major cities before and after sea level increases of 1 to 6 m.

FIGURE I 5-2 A seawall surrounds Malè, the capitol of the Maldives.

Media Analysis 1

Go to www.npr.org, and search for the March 2008 report "The Mystery of Global Warming's Missing Heat."[2] Listen to the 4-minute program, and answer the following questions.

QUESTION 5-3. How much of global warming's heat is going into the oceans?

QUESTION 5-4. What did the 3000-robot "array" of monitoring devices report about ocean warming over the period 2004 to 2008?

QUESTION 5-5. Why does sea level rise when the oceans warm?

QUESTION 5-6. How much has sea level risen over the period 2004–2008?

QUESTION 5-7. Describe the significance of this rise, according to the interviewee.

QUESTION 5-8. How does the observation of sea level rise challenge the data from the robot array?

QUESTION 5-9. Why are clouds called a kind of "natural thermostat?"

QUESTION 5-10. Describe the level of scientific understanding of the role of clouds in climate change over the period of study.

---

2. Or go directly to http://www.npr.org/templates/story/story.php?storyId=88520025.

Media Analysis 2

Go to www.npr.org, and search for the December 2006 program, "Warming Oceans Less Hospitable for Marine Life."[3] Listen to the 3-minute program, and then answer the following questions.

QUESTION 5-11. Why is the recent observation that the ocean is "getting bluer" not good news?

QUESTION 5-12. How do phytoplankton help the planet "breathe?"

QUESTION 5-13. What has the recent change from a greener ocean to a bluer ocean coincided with?

QUESTION 5-14. What are long-term implications for a decline in phytoplankton?

QUESTION 5-15. How does the observed change in phytoplankton affect carbonic acid in seawater, and what are the implications?

---

3. Or go directly to http://www.npr.org/templates/story/story.php?storyId=6591611.

Climate Change and the North Atlantic Circulation

KEY QUESTIONS

What is thermohaline circulation?

What effect will climate change have on circulation in the North Atlantic?

How might this affect climate of northern Europe?

Introduction

We discussed aspects of ocean circulation in Chapter 6. Here we use some of that information to discuss the impact of potential changes in the circulation of the North Atlantic Ocean on the climate of Europe.

In a number of scientific papers in such journals as *Nature*, *Science*, and *GSA Today*, Columbia University Geoscientist W.S. Broecker has warned of "an oceanic flip-flop." Global warming, he argues, could interrupt a system we described in Chapter 6 called thermohaline circulation (also known as meridional overturning circulation). Thermohaline circulation occurs when dense, salty water forms over less dense deeper water, causing the surface water to sink. The entire circulation system (gyre) in the North Atlantic seems to depend on rapid, intense cooling of hypersaline (extra salty) water off Greenland. Prevailing cold, westerly winds chill the salty, warm waters of the Gulf Stream at its northern end, increasing the water's density. Sinking of this dense, now cooled and salty water acts like a conveyer belt to pull the water of the Gulf Stream northward. Thermohaline circulation thus helps maintain the flow of the Gulf Stream that warms much of Europe (Figure I 6-1).

If global warming leads to increased rainfall, reduced wind speed, or higher temperatures over the North Atlantic, or to the melting of freshwater glaciers in Greenland

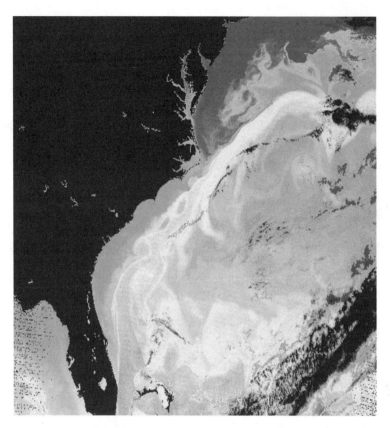

FIGURE I 6-1 False color satellite image of the Gulf Stream off the US East Coast. The Gulf Stream is a powerful warm-water current originating in the tropics. The Gulf Stream plays an important role in the transfer warm and salty water from the tropics poleward. At the latitude of Greenland prevailing westerly winds cool the current by extracting its heat, thus increasing its density, causing its water to sink.

(flooding the area with low-density fresh water), density of surface waters could fall, leading to less mixing of surface and deep waters. This would slow the flow of the Gulf Stream, bringing a cooler, more extreme climate to northern Europe, as the pattern of prevailing winds, if they continue, would blow over a substantially cooler body of water.

More provocatively, Broecker went on to hypothesize that extreme temperature increases could even turn off the deep-ocean conveyor belt completely. Evidence for such flip-flops has been found in geological records obtained from ice cores and deep-sea sediments. Of particular concern is the fact that these events have occurred over time periods as short as 4 years. Broecker refers to the oceans as the "Achilles' heel" of the climate system.[1]

---

1. UN Environmental Programme. GEO 2000: Global Environmental Outlook. Available at http://www.unep.org/Geo2000/.

QUESTION 6-1. In his 1999 article in *GSA TODAY*,[2] Broecker says, "An extreme scenario is an unlikely one, for [present] models suggest that in order to force a conveyor shutdown, Earth would have to undergo a 4 to 5 degree C warming." Eight years later, a conclusion of the 2007 IPCC Report on the impact of global climate change stated that "the amount of global average surface warming following a doubling of carbon dioxide concentrations by 2100 is likely to be in the range of 2 to 4.5°C." Based on this statement, comment on the possibility of a conveyor shutdown in the North Atlantic before 2100. Cite evidence for your conclusion.

Media Analysis 1

Go to www.npr.org, and search for the November 2005 program "Atlantic Ocean's Heat Engine Chills Down."[3] Listen to the nearly 4-minute report. Then answer the following questions.

QUESTION 6-2. How does the Gulf Stream warm Western Europe?

QUESTION 6-3. What do computer forecasts suggest would happen if the Gulf Stream "conveyor" were to stop?

QUESTION 6-4. Describe the controversy over this report in *Nature* that suggested the Gulf Stream might be slowing down.

Media Analysis 2

Go to www.npr.org, and search for "New Data on Earth's Climate History"[4] from 2006 and listen to the 16-minute program. Then answer the following questions.

QUESTION 6-5. What were the striking findings about the Earth's climate 55 million years ago discovered by the ocean drilling expedition to the Arctic?

QUESTION 6-6. Scientists can deduce the temperature of surface water in the past by analyzing the oxygen isotope ratios in shells of planktonic organisms living in the surface water at that time and now preserved in deep ocean sediments. What was the temperature of surface water in the Arctic 55 million years ago?

QUESTION 6-7. What did the commentator mean when he described present climate models as "overly conservative?"

---

2. Broecker, W. S. 1999. What if the conveyor were to shut down? *GSA TODAY* 9:2–7.
3. Or go directly to http://www.npr.org/templates/story/story.php?storyId=5033329.
4. Or go directly to http://www.npr.org/templates/story/story.php?storyId=5447575.

QUESTION 6-8. What was the nature and importance of the "one little rock in particular" as described by Dr. Moran?

QUESTION 6-9. What do the researchers mean when they refer to "greenhouse worlds" and "icehouse worlds?"

QUESTION 6-10. How did Dr. Moran describe the importance of Arctic sea ice to the global climate system?

Is Global Warming Causing More Intense and Frequent Hurricanes?

KEY QUESTIONS

Has the cumulative yearly damage from hurricanes gotten worse over the years as human populations have increased?

Are hurricanes and other tropical storms occurring more frequently?

Has the incidence of larger hurricanes increased with time?

Are hurricanes and tropical storms cyclical in occurrence, and can their severity be predicted?

Can we conclude that global climate change is leading to more intense tropical storms, as some models forecast?

Most of us are sadly familiar with the unprecedented death and destruction caused by 2005's Hurricane Katrina on New Orleans and the Mississippi Coast. The impacts of 1999's Hurricane Floyd, however, also caused severe destruction and ecological damage. On September 15, 1999, only 10 days after tropical storm Dennis had pounded and saturated the same coast, Hurricane Floyd belted the North Carolina coast and interior with 130-mph winds. Previously, it had inundated the administrative capital, Governor's Harbour, of Eleuthera Island, Bahamas. The town did not restore water and power for 5 weeks.

Floyd was no ordinary hurricane. At nearly 1,000 km in diameter, it was nearly twice the size of a "typical" Category 3 (110- to 130-mph winds at center) storm, and the intensity of its winds was also noteworthy—tropical storm-force velocities (40 to 73 mph) ranged across the entire width of that great maelstrom.

Dennis had already saturated the ground, and thus there was no place for Floyd's torrential rains to go. Three hundred years ago, much of the force of the rain would have been spent against the leaves of the region's great old-growth hardwood forests, and the forest soils would have absorbed some of the precipitation, mitigating runoff. Now, however, development and agriculture have laid much of the surface bare, and the rains poured off the region's roads, roofs, parking lots, and plowed fields, carrying untold volumes of soil and surface contaminants with it. Man-made "lagoons" filled with hog excrement overflowed into the rivers—North Carolina by 1999 had more pigs than people. This waste, when mixed with human sewage from washed-out treatment plants, dead animals, agricultural chemicals and fertilizer, and petrochemical sludge from cars and roadways, comprised a "witches' brew" of toxics from which the impacts on the coastal environments and Pamlico Sound are only now being realistically evaluated.

The environmental impacts of hurricanes (Figure I 7-1) can be summarized as follows:

- High winds that can destroy structures, create lethal airborne debris, blow over trees, and so forth
- Storm surges that can push huge volumes of water ashore in the face of powerful winds, especially if coupled with high tides
- Tornadoes spawned by the furious winds
- Inland flooding and landslides

FIGURE I 7-1 Satellite image of Hurricane "Mitch", October 1998 approaching Honduras. The storm devastated the small country's economy, already battered by a long civil war. How can environmental disasters like Mitch contribute to human global migration patterns? Based on climate change models and coastal population growth, are there likely to be more, or fewer, environmental refugees by 2100?

- Pollution surges from animal feedlots, manure and waste "lagoons," agricultural pollutants (fertilizer, pesticides, fuels, etc.), human sewage from flooded treatment plants, toxic chemicals from machinery, industry, autos, and so forth

- Loss of human life and livestock

Many computer models of global climate change forecast a greater incidence of hurricanes, more powerful hurricanes, and/or an increase in the length of the hurricane "season" (which is at present from June 1 to November 30) arising out of increased heating of tropical waters. Is there evidence that hurricane numbers or strengths are increasing? In this brief issue, we assess the evidence that hurricanes are actually increasing in number and/or severity.

What Is a Hurricane?

Hurricanes (also called typhoons or tropical cyclones in some parts of the world) are intense weather systems that develop in the tropics and are defined to have sustained wind speeds of 74 mph or higher.

For a hurricane to form, the following conditions are required:

- Sustained sea surface temperatures of 80°F (26.5°C) or higher (Figure I 7-2)

- Moist air

- Low wind shear

- Coriolis force

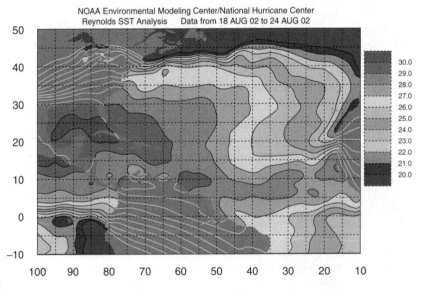

FIGURE I 7-2 Average sea-surface temperature for the North Atlantic for 18 August 2002 to 24 August 2002. Outline the area within which the sea surface temperature exceeds 27°C, the temperature needed to generate hurricanes. Outline the area where hurricanes may form. [*Source:* NOAA.]

Hurricane formation is essentially restricted to the tropics although hurricanes do not form at the equator. Why? There is no Coriolis force at the Equator. Hurricanes basically are the result of intense turbulence formed from convecting air. Heat stored in the ocean is released to the atmosphere when the seawater evaporates, rises, cools, and condenses.

Hurricanes are classified according to the Saffir-Simpson scale (Table I 7-1). They are categorized as major if they reach categories 3 or higher.

Hurricanes and Climate Change

Storm- or hurricane-induced erosion, property damage, and loss of life capture our immediate attention. Better weather forecasting, however, has resulted in a decrease in the loss of life from storms over the last century in the United States.

TABLE I 7-1   The Saffir-Simpson Hurricane Scale. Might climate change require a category 6 to be added to this scale?

- **Tropical Storm**
  Winds 39–73 mph

- **Category 1 Hurricane—winds 74–95 mph (64–82 kt)**
  No real damage to buildings. Damage to unanchored mobile homes. Some damage to poorly constructed signs. Also, some coastal flooding and minor pier damage.
  - Examples: Irene 1999 and Allison 1995

- **Category 2 Hurricane—winds 96–110 mph (83–95 kt)**
  Some damage to building roofs, doors and windows. Considerable damage to mobile homes. Flooding damages piers and small craft in unprotected moorings may break their moorings. Some trees blown down.
  - Examples: Bonnie 1998, Georges(FL & LA) 1998 and Gloria 1985

- **Category 3 Hurricane—winds 111–130 mph (96–113 kt)**
  Some structural damage to small residences and utility buildings. Large trees blown down. Mobile homes and poorly built signs destroyed. Flooding near the coast destroys smaller structures with larger structures damaged by floating debris. Terrain may be flooded well inland.
  - Examples: Keith 2000, Fran 1996, Opal 1995, Alicia 1983 and Betsy 1965

- **Category 4 Hurricane—winds 131–155 mph (114–135 kt)**
  More extensive curtainwall failures with some complete roof structure failure on small residences. Major erosion of beach areas. Terrain may be flooded well inland.
  - Examples: Hugo 1989 and Donna 1960

- **Category 5 Hurricane—winds 156 mph and up (135+ kt)**
  Complete roof failure on many residences and industrial buildings. Some complete building failures with small utility buildings blown over or away. Flooding causes major damage to lower floors of all structures near the shoreline. Massive evacuation of residential areas may be required.
  - Examples: Andrew(FL) 1992, Camille 1969 and Labor Day 1935

Growth in population in hurricane-prone areas, the inflationary increase in property values, and the associated expansion in infrastructure and housing all ensure that catastrophic hurricanes will cause ever-increasing damage in dollar terms, even without the projected effects of climate change. Add the impacts of climate change and more and more Katrinas loom on the horizon. Where, however, is the concrete evidence that climate change has already started to affect severity or number of hurricanes? To assess this question, go to Media Analysis 1.

Media Analysis 1

Go to www.npr.org. Search for "Study: Severe Hurricanes Increasingly Common,"[1] and listen to the 4-minute program from 2005. Then answer the following questions.

QUESTION 7-1. Scientist Peter Webster at Georgia Tech studied hurricanes globally over the 35 years between 1970 and 2005. What were his conclusions?

QUESTION 7-2. What was his conclusion about the incidence of Category 4 and Category 5 hurricanes since 1995 compared with the 1970s?

QUESTION 7-3. What reason did Webster offer for the increase in severity of hurricanes?

QUESTION 7-4. What was scientist Roger Pelkey's response?

Media Analysis 2

Go to www.npr.org, and search for "Insurers Try to Calculate Risks of Climate Change."[2] Listen to the 6-minute program from 2008, and then answer the following questions.

QUESTION 7-5. According to the report, how was the insurance industry responding to the increasing damage from hurricanes in New Orleans and elsewhere on the East and Gulf Coasts?

QUESTION 7-6. What three factors did Risk Management Solutions' representative describe as having contributed to the increased damage caused by hurricanes in New Orleans?

QUESTION 7-7. What "important message" is included in the new higher rates that insurance companies are charging in New Orleans and other risk-prone areas?

QUESTION 7-8. Why are new houses in the Lakeview neighborhood in New Orleans getting higher?

---

1. Or go directly to http://www.npr.org/templates/story/story.php?storyId=4850542.
2. Or go directly to http://www.npr.org/templates/story/story.php?storyId=18288195.

Media Analysis 3

Go to www.npr.org. Search for "Climate Change Fuels Debate over Hurricane Threat,"[3] and listen to the 4-minute program from 2008. Then answer the following questions.

QUESTION 7-9. Describe the nature of the consensus among scientists at the American Meteorological Society meeting on the impact of climate change on hurricanes.

QUESTION 7-10. How did the two contending scientists, Greg Holland of the National Center for Atmospheric Research and Chris Landsey of NOAA, use different kinds of data to support their positions?

QUESTION 7-11. How did Hong Kong scientist Johnny Chan describe his view of the responsibility of climate scientists?

Postscript

On May 3–4, 2008, Cyclone Nargis, a Category 3 storm, hit Burma's coast and Irrawaddy Delta, where 3.5 million of the nation's 50+ million live. The capital, Rangoon (Yangon), was described by survivors as "totally devastated." More than 20,000 people may have been killed, and millions were without food at a time when world rice stocks were low and prices were rising. The population growth rate is approximately 0.8% per year.

QUESTION 7-12. How many more Burmese will inhabit the country in 2009 compared with 2008?

QUESTION 7-13. Comment on the impact of an enhanced cyclone season potentially brought about by global warming coupled with an increase of 400,000+ people per year.

QUESTION 7-14. The nation's geography is described as rugged highlands surrounding a central lowland. Comment on how this could lead to devastating floods during cyclone season.

---

3. Or go directly to http://www.npr.org/templates/story/story.php?storyId=18292843.

PART 5

Marine Pollution

Moms and POPs: Toxic Chemicals in Seawater

KEY QUESTIONS

What are the effects of POPs on marine organisms?

How can we measure these threats?

What can we do about them and who is responsible for doing it?

In this Issue, we examine the impacts of some toxic chemicals on the marine environment.

Since the early 20th century, humans have been synthesizing chemicals not naturally found on earth, chemicals that microbial "detoxifying" agents such as bacteria seemingly have limited capacity to render harmless. Concentrations of many of these chemicals have reached levels in the oceans that have begun to threaten the survival of many species of phytoplankton, invertebrates, large mammals, and fish. Because these chemicals tend to be difficult for organisms to decompose, they are very persistent in seawater. Moreover, even if organisms can break the chemicals into "byproducts," these decomposition products are often toxic as well. DDE, from DDT, is one notorious example. We briefly list and describe the most important of these chemicals, called POPs, or persistent organic pollutants, and describe their effects on marine food webs and chains. Then we describe efforts to eliminate them.

POPs

POPs include a bewildering array of synthetic organic compounds that (1) persist in the marine environment, (2) are easily distributed globally, and (3) lodge and become concentrated in the fatty tissues of animals by bioaccumulation and biomagnification, processes we introduced in Chapter 6.

TABLE I 8-1	The Dirty Dozen and Their Origins

The "Dirty Dozen"

[1=Pesticide]

Chemical	Comments
aldrin[1]	Fatal dose, 5g, Adult male: fatal dose for women and children less.
hexachlorobenzene[1]	Can be lethal: has been found in food of all types.
chlordane[1]	Toxic to many animals. Human exposure is mainly by air.
mirex[1]	Used against fire ants. Very stable and persistent. Possible human carcinogen.
DDT[1]	Has been detected in breast milk. May harm infants.
toxaphene[1]	Most widely used pesticide in the US in 1975. Possible human carcinogen.
dieldrin[1]	Mutagenic. Second most common pesticide detected in a US survey of pasteurized milk.
polychlorinated biphenyls (PCBs)	Suppress human immune system; probable human carcinogen; readily transferred in breast milk.
endrin[1]	Toxic, but can be metabolized, so little bioaccumulation.
polychlorinated dibenzo-p-dioxins (dioxins)	Seven types out of 75 are mutagenic, carcinogenic.
heptachlor[1]	High doses fatal to birds, mammals; low doses mutagenic.
polychlorinated dibenzo-p-furan (furans)	135 different types; possible human carcinogen; can accumulate in breast milk.

Source: EPA and World Bank.

■ POPs: The Dirty Dozen

The EPA calls the 12 POPs in Table I 8-1 the "dirty dozen." Most are pesticides, part of the "Green Revolution" that radically increased food production during the 20th century. Dioxins and furans are industrial byproducts of combustion. Polychlorinated biphenyls (PCBs) were used for years in electrical insulation. The manufacture and use of all POPs (except for furans and dioxin, which are combustion by-products) have for years been banned in the United States. All are easily biomagnified, as shown in Table I 8-2.

QUESTION 8-1. By what factor is toxaphene concentrated in zooplankton as compared with seawater?

QUESTION 8-2. By what factor is toxaphene concentrated in seal blubber compared with its food, arctic char?

TABLE I 8-2	**Biomagnification of the POP toxaphene (Adapted from http://www. biology.duke.edu/bio217/2002/pcb/Toxaphene.htm)**

Toxaphene Biomagnification - Arctic Canada
Compartment Concentration

Air .0007 ppb
Snow .0009–.002 ppb
Seawater .0003 ppb

Zooplankton 3.6 ppb
Arctic cod muscle 14–46 ppb
Arctic char whole body 44–157 ppb

Ringed seal blubber 130–480 ppb
Beluga blubber 1380–5780 ppb
Narwhal blubber 2440–9160 ppb

Chart information compiled by Lars-Otto Reiersen.

QUESTION 8-3. Ringed seals are preyed upon by beluga whales. But what factor is toxaphene concentrated in beluga blubber compared with seal blubber?

■ PCBs

PCBs (Figure I 8-1) are a group of over 200 synthetic chemicals called organochlorines, which are compounds made of one or more chlorine (Cl) atoms attached to carbon-based structures such as chains and rings. Their formulas are complex, and their atomic weight depends on the number of relatively heavy chlorine atoms in the structure.

Let us break down the name to explain what kind of chemical we are dealing with. First, "poly" means "many;" "chlorinated" refers to chlorines that are attached to the carbon atoms; and "phenyl" refers to a type of organic structure with six linked carbon atoms arranged to form a hexagonal ring. The phenyl ring may be attached to more than one chlorine atom, whence the phrase "polychlorinated."

QUESTION 8-4. How many phenyl rings are linked in PCB (hint: what does the prefix "bi" mean)?

PCBs do not exist naturally on earth: They were synthesized during the late 19th century. Because of their stability when heated, they were widely used in electrical capacitors and transformers to prevent electrical fires. In the 1960s, scientists began to report toxic effects on organisms exposed to PCBs, and by 1977, the manufacture of PCBs was banned in the United States, the United Kingdom, and elsewhere. By 1992, scientists estimated that 1.2 million metric tons of PCBs existed worldwide, while

FIGURE 18-1 Structure of a PCB. Note that there are 2 hexagonal phenyl groups, each with 5 Cl.

scientists estimate that 370,000 metric tons (810 million pounds) have been dispersed globally, much of it into the oceans.

PCBs are relatively insoluble in water as they are lipophilic (meaning that they are very fat soluble) and thus tend to accumulate in the fatty tissues of animals (bioaccumulation). Concentrations in seawater may reach 1 part per million (ppm), but PCBs typically concentrate in sediment. From there, they enter food chains and webs (collectively called "trophic tiers"), mainly through the feeding of organisms called sediment or deposit feeders, discussed in Chapter 6. These animals eat sediment, extract organic matter, and excrete the rest. If other animals eat the deposit feeders, the PCBs move up the trophic tiers and become further concentrated. Concentrations exceeding 800 ppm have been measured in the tissues of large marine mammals. According to the Environmental Research Foundation, this would qualify the creature for hazardous waste status!

PCBs have become widespread and serious pollutants and have contaminated many marine trophic tiers. They are extremely resistant to breakdown (which of course was one of their virtues) and are known to be carcinogenic and probably mutagenic (mutation inducing) as well. In terms of their specific effect on life, PCBs have been shown to cause liver cancer and harmful genetic mutations in animals. PCBs have been linked to mass mortalities of striped dolphins in the Mediterranean, to declines in Orca (killer whale) populations in Puget Sound, and to declines of seal populations in the Baltic.

Although PCBs may threaten the entire ocean, the northwest Atlantic is believed to be the largest PCB reservoir in the world because of the amount of PCBs produced in countries that border the north Atlantic.

According to a report edited by Paul Johnston and Isabel McCrea for Greenpeace UK,

Since the rate at which organochlorines break down to harmless substances (has been) far outstripped by their rate of production, the load on the environment is growing each year. Organochlorines (including PCBs) are arguably the most damaging group of chemicals to which natural systems can be exposed.

PCBs and Orcas in Puget Sound

Orcas, large-toothed whales, are the largest members of the dolphin family of cetaceans. Although soldiers during World War II used them for target practice, Orcas have since become a symbol of the Pacific Northwest. Figure I 8-2 shows a killer whale (Orca) from Puget Sound, Washington.

In 1999, Dr. P.S. Ross,[1] a research scientist with British Columbia's Institute of Ocean Sciences, took blubber samples from 47 live killer whales and found PCB concentrations from 46 ppm to over 250 ppm, up to 500 times greater than concentrations found in humans. Ross concluded, "The levels are high enough to represent a tangible risk to these animals."

Ross compared the Orca population he studied with the endangered beluga whale population of the St. Lawrence estuary of eastern North America, in which a high incidence of diseases involving compromised immune systems has been linked to contaminants and which has shown evidence of reproductive impairment.

In Orcas, the PCBs are passed from generation to generation. PCBs are highly fat soluble and thus become concentrated in mother's milk. Ross said, "Calves are bathed in PCB-laden milk at a time when their organ systems are developing and they are at their most sensitive." In 2007, Ross and colleagues reported that, because of biomagnification and bioaccumulation, Orcas could face toxicity hazards for up to 60 years after production of PCBs was halted.

FIGURE I 8-2 Orcas in Puget Sound, Washington. One Puget Sound group is a "listed" species under the U.S. Endangered Species Act.

1. Ross, P. S., G. M. Ellis, M. G. Ikonomou, L. G. Barrett-Lennard and R. F. Addison. 2000. High PCB concentrations in free-ranging Pacific killer whales, *Orcinus orca*: effects of age, sex and dietary preference. *Marine Pollution Bulletin.* 40: 504–515.

Although PCBs have been banned in the United States for 20 years, they are still being used in many developing countries. Approximately 15% of known PCBs reside in developing countries, mostly as a result of shipments from industrialized countries. Accordingly, Ross speculates that PCBs in the Pacific could be derived from East Asian sources and could end up concentrated in the tissues of migratory salmon, which are a prime food source for the Orcas.

J. Cummins, in a 1988 paper in *The Ecologist*, stated that adding 15% of the remaining stock of PCBs to the ocean would result in the extinction of marine mammals.

QUESTION 8-5. If PCBs are only slightly soluble in seawater, how can adding a small percentage of remaining PCBs to the ocean threaten the extinction of all marine mammals?

In this analysis, you will examine the impact of the PCB dioxin, which is, according to the U.S. EPA, "one of the most toxic substances known." How long PCBs persist in the body of an animal is variable, but PCBs are not easily metabolized and excreted. Estimates from EPA range from 1 to 10 years.

QUESTION 8-6. Estimate how long it may take for PCB concentrations to fall by one half, assuming that no new PCBs are added to the system.

QUESTION 8-7. Marine mammals are particularly susceptible to the harmful reproductive effects of PCBs. Why is biomagnification potentially so dangerous?

QUESTION 8-8. Recall the concept of a food chain from Chapter 7, as a concept that ties primary producers to herbivores, first-level carnivores, and so on. Which type food chain, a "short" one with few levels or a long one with many levels, would be more susceptible to the buildup of dangerous levels of POPs? Formulate a hypothesis to test your conclusion.

QUESTION 8-9. Some readers may conclude that Greenpeace scientists, one of our sources for this issue, may be too "biased" to provide factual data on PCBs. Research PCBs on the Internet or in an environmental science textbook. Describe any differences in the treatment of the PCB threat you observe.

What Can We Do About POPs and Who Is Responsible?

In 2001, U.S. representatives signed the Stockholm Convention on POPs, joining 121 other nations. The intent of the treaty was ambitious: not to simply control POPs but to eliminate and destroy them. The treaty has the following elements:

1. Funding commitments from developed countries to developing countries, to assist them in eliminating POPs.

2. Eliminating intentionally produced POPs. DDT was recognized as a special case because of its short-term effectiveness in controlling malaria.

3. Ultimate elimination of "by-product" POPs.

   QUESTION 8-10. Which members of the Dirty Dozen were by-products?

4. Environmentally sound disposal or destruction of POPs.

5. Open exchange of information about POPs: specifically, claims of confidentiality were to be disallowed due to the overwhelming health risk.

6. Strict controls on global transport of POPs, generally to be allowed only for destruction or disposal.

7. Ultimate addition of other hazardous chemicals to the Dirty Dozen list.

The greatest challenge for the present is to end practices that produce by-product POPs, such as the production of dioxin during the bleaching of paper by chlorination.

Media Analysis: Human Impact on the Oceans

For a global perspective on humanity's cumulative impact on the oceans, go to www.npr.org, and search for "Scientists Map Ocean Damage."[2] Listen to the 4:23 audio. Then click on the animated global map, and view the 2-minute animation. Answer the following questions.

QUESTION 8-11. In what journal was the map published? If this map was published by Greenpeace, would readers give it equal regard? Why or why not?

QUESTION 8-12. What do "red zones" represent?

QUESTION 8-13. Where are the reddest of the red zones located?

QUESTION 8-14. Characterize the general health of the polar oceans.

QUESTION 8-15. Given the projection that climate change is due to affect the poles much more severely than other latitudes, speculate on the "business-as-usual" future of polar oceans.

QUESTION 8-16. The Persian Gulf is one of the most degraded zones in the oceans. Give reasons for why this is so.

QUESTION 8-17. Where are the most disturbed zones of coastal North America? Explain why this is the case.

---

2. Or go directly to http://www.npr.org/templates/story/story.php?storyId=19059595.

Just Cruisin': The Environmental Impact of "Cities at Sea" and Cargo Ships

What are the environmental impacts of cruise ships?

What are the environmental impacts of cargo ships?

Which ships, cruise or cargo, have the greatest global impact?

What can be done to lessen these impacts?

Some of these ships are huge, like great power stations on the waves, and they can produce vast amounts of sulfur and nitrogen oxides, both of which contribute to acid precipitation.

—Professor David Fowler of the Center for Ecology and Hydrology in Edinburgh, Scotland[1]

Every one of Carnival's "Fun Ships" is a unique floating resort designed with your fun in mind. Venture out of your spacious stateroom and experience the outdoor areas, wonderful restaurants, friendly casino, relaxing lounges, invigorating spa, exciting nightclubs and duty-free shopping. Come aboard and see for yourself.

—From Carnival Cruise Line's website (www.Carnival.com)

Picture this "fun ship" with its "spacious stateroom, … wonderful restaurants, friendly casino" and a huge banner reading "Got Sewage?" flying high above the ship (Figure I 9-1). You might react this way: Sewage? This ship dumps its sewage at sea?

---

1. Ships sabotage war on acid rain. *The Observer*, UK, Sunday October 10, 2004.

Actually, it is perfectly legal for cruise ships to dump treated sewage virtually anywhere they cruise and to dump untreated sewage outside of the United States 12-mile jurisdictional limits. The sign above the ship in Figure I 9-1 refers to illegal dumping of raw or under-treated sewage as well as oil-contaminated waste by Royal

FIGURE I 9-1 "Got Sewage" banner flying over cruise ship.

Caribbean Cruises Ltd., for which the company pled guilty to multiple felony counts and was fined more than $26 million in 1998 and 1999.

According to the General Accounting Office, from 1999 to 2005, cruise ship lines have paid over $41 million in fines for illegally discharging oily waste and toxic chemicals (including dry cleaning chemicals) and dumping plastic bags of garbage into the ocean.

Unfortunately, these are not the only environmental problems associated with cruise ships. Moreover, pollution from cargo ships may be even more hazardous.

Impact of Cruise Ships

Royal Caribbean Line's Voyager-class vessels are the largest of their kind in the world. The *Mariner* is over 1,000 feet long, displaces 138,000 tons, and can accommodate more than 3,800 passengers (plus an additional 1,200 crew members). It even has an ice skating rink. *Mariner* cruises throughout the Caribbean at a speed of 22 knots, powered by diesel-electric engines producing 75,600 kW (75.6 MW). Like all cruise ships, *Mariner* has considerable environmental impact (Table I 9-1). Globally, as many

---

TABLE I 9-1   Enviornmental Impact from the Cruise Ship Mariner

- **Sewage**
  >140,000 gallons daily. "Graywater" from laundry, showers, sinks, dishwashers, etc.

- **Solid waste**
  Nearly 12 tons of solid waste and garbage a day.

- **Toxic chemicals**
  5 gallons a day, including silver, turpentine, mercury, lead, and cadmium from photo processing, dry cleaning, painting, etc.

- **Oily waste**
  Approximately 4,000 gallons, including residual oil from routine engine maintenance produced daily.

- **Ballast water**
  Hundreds of thousands of gallons of seawater used to stabilize the vessel for the duration of a cruise. Ballast water can transport alien species, like Chesapeake Bay's veined rapa whelk. (See Issue 13, Illegal Immigration)

- **Air pollution**
  A typical cruise ship produces: >1.5 tons of smog-forming nitrogen oxides, >1.3 tons of sulfur oxides, >100 pounds of volatile organic compounds, >75 pounds of particulate matter, and carbon dioxide daily.

as 200 cruise ships carry over 10 million passengers annually, and that number is projected to increase.

Some would add to the items in Table I 9-1 the off-ship impact, which is simply the sum of all of the waste- and pollution-generating activities that passengers on and suppliers to *Mariner* produce before they even arrive at the ship. For example, many people fly to their port of embarkation, resulting in added pollution from the burning of jet fuel.

QUESTION 9-1. Do you agree? Why or why not?

Impact of Cargo Ships

Studies published at the end of 2004 demonstrated that air pollution from seagoing cargo ships, including oil tankers, was a significant and growing source of dangerous coastal air pollution.

Moreover, unlike other pollution sources such as refineries, motor vehicles, and even wood stoves, authorities in the United States, whether local, state, or federal, have limited authority to address this problem. For example, a 1944 treaty prohibits taxing fuel for international transport of goods.

QUESTION 9-2. Does the 1944 treaty provide a justifiable subsidy favoring cargo ships, which use untaxed fuel, over trucks that carry goods from country to country, using fuel taxed in the country of origin? Why or why not?

The U.S. Environmental Protection Agency (EPA) has begun regulating emissions from American ships, although its rules do not cover foreign-flagged vessels that make up more than 90% of the ships calling on some West Coast ports. (The maritime industry opposed the EPA's attempts to set limits on air pollution from large ships, arguing that current information didn't support the move.)

The Significance of Marine Sources of Pollution

Here are examples of the impact of marine pollution:

- A 2003 British Columbia study concluded that marine vessels produced more than a million tons of pollutants a year, more than half the total measured annually. An official concluded in 2004 that "international foreign-flag ships, most of which burn high-sulfur bunker oil, are the region's major source of dangerous sulfur emissions."

- According to Seattle, WA officials, each day a freighter sits in Elliott Bay, the city's harbor, the ship's smokestacks can emit as much nitrogen oxide into the city's air as 12,500 cars, or as much as an oil refinery each day. Moreover, The Puget

Sound Clean Air Agency has estimated that diesel-powered vessels plying the Sound produce 15% of total sulfur oxides (SOx) and 7% of total fine particulates. The agency projected the engines will produce 25% of mobile-source emissions by 2020.

- In the Los Angeles and Long Beach harbors, a typical oceangoing ship produces as much nitrogen oxide in a day, 2 to 3 tons, as an oil refinery.

- In Britain, monitoring has documented a decline in emissions of sulfur dioxide and nitrogen oxides (NOx) from factories and power stations; however, increases in SOx, NOx, and particulates from tankers and container ships have negated many of the gains. For the whole of Europe, marine sources of SOx accounted for 10% of the acid rain in 1990, but by 2010, they will be responsible for 45%. Shipping in the Baltic, North Sea, and North Atlantic accounted for 2 million tonnes of SOx in 1990 and increased to 2.6 million tonnes in 2000. By 2010, they are expected to contribute 3.3 million tonnes.

- The EPA lists several cities with significant marine-derived air pollution as "nonattainment areas" in terms of air pollution.

- The fuel used by most tankers, called "bunker fuel," has the highest sulfur content of any petroleum-based fuel. Although new regulations in 2006 reduced the sulfur content of diesel fuel burned by on-road vehicles in the United States to 15 parts per million (ppm), the sulfur content of bunker fuel can be as high as 4.5%.

QUESTION 9-3. Convert 4.5% to parts per million (hint: a 1% concentration is equal to 10,000 ppm).

QUESTION 9-4. By what proportion does the sulfur in bunker fuel exceed the sulfur content in on-road diesel?

QUESTION 9-5. At a sulfur content of 4.5%, how much sulfur, by weight, is contained in a tonne of bunker fuel (one tonne = 1,000 kg)?

QUESTION 9-6. The atomic weight of sulfur is 32, and that of oxygen is 16. What is the combined atomic weight of SO_2, which is produced when sulfur is burned along with the bunker fuel?

Thus, each atom of S combines with an equal mass of oxygen to produce SO_2. **QUESTION 9-7.** How much SO_2 is produced when a tonne of bunker fuel is burned?

Some scientists believe high-sulfur fuels are contributing significantly to global warming. Some of the black soot (particulates) settles out of the atmosphere and covers ice at the poles, darkening it and increasing its tendency to melt.

QUESTION 9-8. Would you favor an international tax on bunker fuel to compensate for its environmental impact or even banning it outright? Why or why not?

QUESTION 9-9. How does the use of cheap, polluting, untaxed bunker fuel by marine cargo vessels contribute to the loss of jobs in developed countries as manufacturing shifts to developing countries?

QUESTION 9-10. Some port cities like Gothenburg, Sweden and Long Beach, CA have begun to offer cargo ships the opportunity to hook into electricity supplies on shore so that they can turn off their engines while loading and unloading freight. Would you favor international regulations to make this a requirement rather than an option? Why or why not?

Media Analysis 1

Go to www.npr.org, and search for "cruise pollution."[2] Listen to the 7-minute May 2001 program entitled "Cruise Pollution," and then answer the following questions. (By 2008, the Supreme Court had agreed that the U.S. government had authority to regulate cruise ships.)

QUESTION 9-11. Describe the nature of "black water" and "grey water" discharged from cruise ships into the coastal waters of Alaska.

QUESTION 9-12. What proportion of sewage samples taken from the cruise ship met standards?

QUESTION 9-13. How did cruise ships as of 2001 avoid regulation?

QUESTION 9-14. How much money did cruise ships bring to southern Alaska communities as of 2001?

QUESTION 9-15. What tax was to be imposed to pay for Alaska's proposed regulation?

QUESTION 9-16. How much were cruise ships set to invest per ship to meet the proposed Alaska regulations?

QUESTION 9-17. At the end of the program, what did the Coast Guard cite the first cruise ship to leave Juneau for?

Media Analysis 2

Go to http://savvytraveler.publicradio.org/show/features/1999/19990116/cruise.shtml, and listen to the 2-minute report on cruise ship pollution.

QUESTION 9-18. What is the most common pollution problem linked to cruise ships according to Captain Powers?

---

2. Or go directly to http://www.npr.org/templates/story/story.php?storyId=1122792.

QUESTION 9-19. How are offenses most often detected?

QUESTION 9-20. What does Powers recommend that concerned cruise ship passengers do to record dumping abuses by cruise lines?

QUESTION 9-21. Would this kind of environmental impact affect your decision on whether to take a cruise? Why or why not? Assess your response in terms of critical thinking. Did your first impressions contain any logical fallacies?

Human Impacts on Estuaries

KEY QUESTIONS

What are estuaries and how are they formed?

What are the ecological functions of estuaries?

Why is biological productivity high in estuaries?

Why are sustainable estuaries important to human society?

What human activities adversely impact estuaries?

What can be done to avoid or lessen these impacts?

Introduction: Ecological Function of Estuaries

Estuaries (Figure I 10-1) are coastal embayments where fresh and marine waters mix. Most of today's estuaries are all less than 12,000 years old, resulting from the geologically recent rise of sea level since the last glacial maximum.

The shallow, nutrient-rich waters of estuaries, along with the intertidal wetlands that line their shores, are some of the most productive ecosystems on earth. Estuaries support abundant grasses and macroalgae and in tropical environments, mangroves. These environments serve as nurseries for the juvenile stages of commercially valuable fish and shellfish. Figure I 10-2 shows an estuarine food web.

Even though primary production (the production of organic matter by photosynthesis, Chapter 6) is high in estuaries, animal communities are of low diversity, as many species cannot tolerate the intertidal exposure, dehydration, and sometimes-large variations in temperature and salinity that characterize estuaries. The species that can withstand these factors are highly abundant, as you might expect.

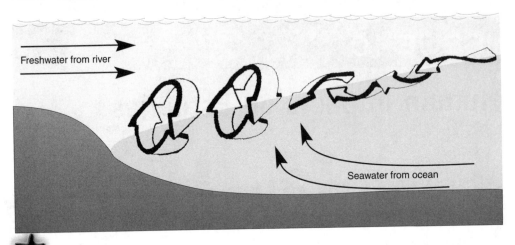

 FIGURE I 10-1 Illustration showing the general pattern of mixing of fresh and saltwater in an estuary.

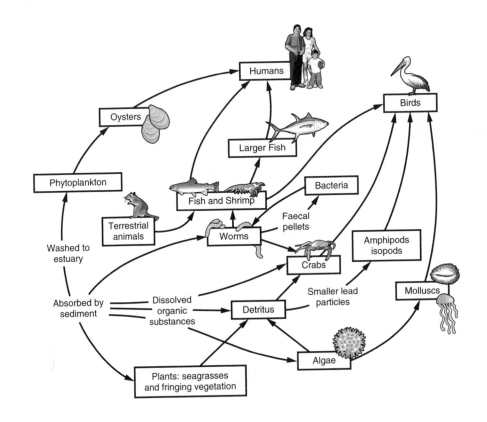

FIGURE I 10-2 An estuarine food web. What do birds feed on in this food web?

Estuaries and the wetlands that fringe them perform valuable ecological services for human society.

- Estuaries provide habitat for more than 75% of America's commercial fish catch and for 80% to 90% of the recreational fish catch.

- Water draining from uplands can carry sediments, nutrients, and pollutants. As the water flows through fresh and salt marshes, some of the sediments and pollutants are filtered out, which benefits both people and marine life.

- Beaches and wetlands surrounding estuaries buffer coastal land and habitats from erosion by the ocean, preventing coastal erosion and habitat destruction.

Population Growth in Coastal Regions

Many large cities in coastal regions are located along estuaries in the U.S., including Boston, Baltimore, San Francisco, and Seattle. Rivers connected to estuaries, when dredged, expand the ability of cities far from the estuary mouth to develop port operations and dependent industry. Floodplains of estuaries and rivers possess rich soils that enhance agricultural production. Having nearby ports for marketing these products encourages agriculture.

Population growth has been much greater in coastal and estuarine regions of the U.S. compared with inland regions (see Figure I 3-1). More than 160 million people—over half the population of the United States—reside along our coasts. This coastal population is increasing by 3,600 people per day.

Human population growth and associated activities within estuarine watersheds have degraded water quality and have altered estuarine ecosystems, as we describe next.

Human Effects on Estuarine Water Quality

- High rates of nutrient input from sewage plants, lawns, agriculture, motor vehicles and fossil-fueled power plants lead to eutrophication and contribute to fish disease, harmful algal blooms, low dissolved oxygen, and change in plankton and benthic community structure.

- Toxic chemicals and pathogens enter estuaries as a result of human activity. Organic substances and metals can have direct, acute effects on the aquatic community. Toxics often bioconcentrate and may reach high concentrations in higher trophic level species (see Issue 8). Human exposure to toxics or pathogens in coastal waters or the consumption of undercooked or raw seafood can cause severe illness or death. Coastal population growth can lead to an increasing concentration of pathogens (viruses, bacteria, and parasites) in coastal waters, primarily from sewage.

- Many human activities degrade estuarine habitat. Examples include conversion of open land for development, cutting of forest, agriculture, installation of dams and culverts, and highway construction.

- Humans can change the hydrology of estuarine watersheds by channelization and damming of rivers, withdrawal of fresh water, and excess heating through power plant emissions, especially relatively inefficient nuclear power plants.

- Dredging channels to promote port operations helps move salty oceanic water into estuaries, changing the salinity structure, circulation, flushing, and water residence times. Moreover, dredged material may be dumped into "spoil areas," burying former healthy communities. Such changes can have serious effects on biological productivity and ecosystem function.

- Resource exploitation can affect ecosystem health. Like many other human activities, fishing can have deleterious ecological effects. Overharvesting of fisheries can cause species collapse. In Chesapeake Bay (Figure I 10-3), the Atlantic oyster (*Crassostrea virginica*) formerly covered much of the subtidal bottom and could filter the entire bay's water volume in a few days. The overharvesting of oysters during the last century has eliminated commercial oyster reefs. The organic matter formerly filtered by oysters now falls to the bottom and is consumed primarily by benthic microbes. This eutrophication contributes to increased frequency, extent, and duration of low oxygen (anoxia and hypoxia) conditions.

- Humans have introduced exotic species, causing major ecological problems. In San Francisco Bay, the dominant benthic organism is the exotic Chinese clam (*Potamocorbula amurensis*), and its filter-feeding activities have eliminated the normal summer phytoplankton blooms in the northern portions of the bay.

  Diseases that are ravaging oyster populations in Chesapeake Bay may have been introduced with oysters transplanted from foreign regions. More recently, the veined rapa whelk (*Rapa venosa*) has been found in Chesapeake Bay (see Issue 22). Its feeding on hard clams represents a threat to the last commercial shellfish resource currently being harvested.

Focus: Chesapeake Bay

In this analysis, we examine the overall health of Chesapeake Bay, the sources and the impacts of pollution, and actions that are being taken to reduce the environmental impact of population growth.

Formed 8,000 years ago as rising sea level drowned the mouth of the Susquehanna River, Chesapeake Bay is the largest and most productive estuary in the U.S (Figure I 10-3). The Bay's watershed, the area drained by streams that feed the Bay, is 64,000 mi^2 (41,000,000 ac) of agricultural land, urban and suburban areas, and forests. It includes parts of six states and the District of Columbia and over 1,650 local governments. The watershed has a human population of about 16 million. Surprisingly, most residents of the Bay's watershed are unaware of its area, and most also do not realize that they live within the watershed. Annually, about 130,000 people move into the watershed.

QUESTION 10-1. What is the human population growth rate for the watershed of the Chesapeake Bay?

FIGURE I 10-3 Chesapeake Bay, the largest estuary in the US. Treated sewage from the watershed's 16 million residents is a major source of contamination. Other sources are air deposition from autos and power plants in the region's airshed, which extends all the way to Indiana and beyond. Should localities that border the Bay be solely responsible for maintaining Bay health?

QUESTION 10-2. Does the population growth rate you just calculated sound large or small?

QUESTION 10-3. Use the population growth rate that you just determined to calculate the doubling time for the population (use 70/r, see Appendix 1 to review the concept of doubling time).

QUESTION 10-4. The Bay's airshed, the geographic area that is the source of airborne pollutants that can affect the Bay, is 6.5 times the area of the watershed. How many square miles is the airshed?

QUESTION 10-5. Discuss whether you think states and Canadian Provinces in the Chesapeake Bay watershed have a responsibility to reduce levels of airborne pollutants that may reach the Bay.

The Environmental Health of the Bay

In 1997, the Chesapeake Bay Program, a regional partnership of local, state, and federal governments, non-profit organizations, academic institutions, and watershed residents working to restore the Bay, published the report *Chesapeake Bay 2007 Health and Restoration Assessment.*[1] This publication lists two major factors impacting the health of the Bay and its watershed: river flow and resultant pollutant loads reaching the bay, and land use.

River Flow and Pollutant Load

The volume of water that enters the Bay from its rivers determines in part how well mixed the Bay is, which in turn impacts the levels of dissolved oxygen, the salinity of the Bay, and the levels of sediment and pollution that reach the Bay. River flows have remained steady in recent years; however, the effect of global climate change will likely change the amount and/or patterns of precipitation in the Bay's watershed.

The Environmental Protection Agency (EPA) lists large areas of the Bay as impaired.[2] In 1983, the EPA concluded that one of the most serious threats was nutrient enrichment, leading to eutrophication. Nutrients are substances that are essential for plant growth, predominantly compounds of nitrogen and phosphorus. An excess of nutrients causes nutrient pollution, which can lead to a massive growth of algae called a *bloom*. Environmental hypoxia (low dissolved oxygen) and anoxia (no dissolved oxygen) can result when the plants die, fall to the bottom, and microbial, aerobic respiration depletes dissolved oxygen (see Issue 17).

In the case of the Bay, however, the algal blooms caused by nutrient pollution not only cause declines in dissolved oxygen, but they also physically block sunlight, depriving one of the Bay's most important inhabitants, sea grasses, of the light the grasses needed to survive.

Until the 1950s, Bay grasses (also called submerged aquatic vegetation) had covered most of the Bay's bottom shallower than 2 meters (6.5 feet), but by 1983, their area had been reduced by 90%. Bay bottom grasses are considered one of the major keys to a healthy Chesapeake Bay. The grasses contribute oxygen to the water, provide food for some Bay organisms, serve as attachment spots for a myriad of tiny organisms, and offer hiding places for juvenile fish, crab larvae, and a host of other small animals (and are vitally important to female blue crabs).

As discussed previously, the result of the blooms was a decrease in the extent of Bay grasses and hypoxic or anoxic water conditions. Lacking plant cover and depleted of oxygen, the bottom becomes a veritable desert, as most bottom-dwelling animals cannot

---

1. Chesapeake Bay Health and Restoration Assessment: A Report to the Citizens of the Bay Region. 2007. Chesapeake Bay Program. Available at http://www.chesapeakebay.net/content/publications/cbp_26038.pdf.
2. http://oaspub.epa.gov/waters/region_rept.control?p_region=3#IMP.

live without oxygen and they die or leave. Thus, the bottom becomes inhospitable to aerobic organisms without Bay grasses to replenish the depleted oxygen and serve its other functions. Other changes may also take place; for example, green algae may be replaced by cyanobacteria and diatoms (for details, see Cooper and Brush[3]).

In 1987, the states of Maryland, Virginia, and Pennsylvania, the District of Columbia and a number of federal agencies formed the Chesapeake Bay Commission to coordinate Bay restoration and protection. One of the Chesapeake Bay Commission's goals was a 40% reduction in phosphorus by the year 2000, a goal that it was unable to meet, although progress had been made. The commission also set annual numerical goals of 8.43 million pounds per year for phosphorus reduction. Phosphorus was reduced by 16% by 1992, mainly as a result of a ban on phosphate-bearing detergents that took effect in January 1989, over the strong objection of detergent manufacturers. In 1992, the average phosphorus content in the Bay was 0.03 mg/L. For comparison, streams that drain forested areas contain around 0.02 mg/L phosphorus and those that drain pasture contain around 0.05 mg/L; a representative value for streams that drain urban areas is 0.075 mg/L, and those that drain farmland contain about 0.15 mg/L phosphorus.

QUESTION 10-6. Calculate the volume of Chesapeake Bay. The Bay's surface area is roughly 11,000 km^2, and the average depth is 7 m. Give your answer in cubic kilometers, cubic meters (there are 10^3 m^3 in 1 km^3), and liters (1 m^3 = 1,000 L).

QUESTION 10-7. How much phosphorus was dissolved in the Bay in 1992? Express your answer in kilograms.

As of 2006, despite nearly 3 decades of study and restoration plans, the Bay was described as an "ecological disaster area" by the Chesapeake Bay Ecological Foundation.

Bacteria levels in the Bay in 2006 were among the highest measured in any estuarine environment in the United States. Zooplankton, the food base for many fish species, were becoming scarce in the Bay's main stem during the summer growing season, in part because of the proliferation of comb jellyfish, a major predator of zooplankton and fish larvae. Algal blooms are becoming more common in the Bay.

The Chesapeake Bay Ecological Foundation lists emissions from coal-fired power plants, motor vehicles, agricultural runoff from farms and concentrated poultry and hog operations, and silt from sprawl development as among the causes for the Bay's decline.

Another cause is sewage. The *Roanoke Times* editorialized in 2006 that the Bay was "just as dirty as it was 20 years ago." It cited the cause to be reluctance on the part of states and the federal government to spend more than a trivial fraction of the amount scientists had estimated was needed to remediate the Bay.

By 2000, sewage flow from all sources in Virginia's portion of Chesapeake Bay's watershed was estimated at 522 million gallons per day (1 gal = 3.8 L).

---

3. Cooper, S. R., & G. S. Brush. 1991. Long-term history of Chesapeake Bay anoxia. *Science* 254:992–996.

QUESTION 10-8. Based on this rate of sewage flow, how long would it take for Chesapeake Bay to fill with treated sewage, assuming that the sewage was not flushed to the ocean?

QUESTION 10-9. As part of Chesapeake Bay protection efforts, Virginia required that all municipal and industrial sewage treatment plants and all industries that emit phosphorus reduce the phosphorus concentration in sewage outflow to a level of 2 mg/L by the year 2000. Based on attainment of that standard, how much phosphorus did Virginia sewage treatment plants discharge into the Bay in 2000? Express your answer in kilograms and pounds per year.

QUESTION 10-10. At 2 mg/L, how does sewage runoff compare to the typical phosphorus values for streams given above?

QUESTION 10-11. Do you think a 2 mg/L standard for P is sufficient to restore the Bay? State your reasons. What additional information would you like to have in order to answer this question more thoroughly?

Recent research suggests that between 20% and 35% of nitrogen in Chesapeake Bay is from air pollution. A third of this air pollution is from cars, power plants, and farm fields in the watershed itself, but as much as two thirds is from power plant emissions in Ohio, Kentucky, Michigan, and other states as far away as Alabama. This is apparently due to three factors:

1. Prevailing winds that blow from the southwest, northwest, and west
2. Nitrogen pollution from midwestern coal-burning utilities
3. Smokestacks from the plants that emit the pollution high enough into the atmosphere so that it can be carried over 800 km before settling out

Accordingly, to address water and air pollution problems in the Bay's watershed and throughout the eastern United States, the U.S. EPA has ordered 22 states in an arc from Massachusetts to Missouri to reduce pollutants, including nitrogen (in the form of oxides of nitrogen), or lose federal highway funds. The affected states had until 2005 to actually reduce emissions. The states most heavily affected and the percentage of nitrogen oxide reductions required are as follows: West Virginia (44%), Ohio (43%), Missouri (43%), Indiana (42%), Kentucky (40%), Illinois (38%), Alabama (36%), Wisconsin (35%), Tennessee (35%), and Georgia (35%).

QUESTION 10-12. To what extent should the people in these states be held accountable for air pollution carried beyond their boundaries? Include your reasons and explain them.

Land Use

Impervious surface, any surface material through which water cannot penetrate (roads, parking lots, and rooftops), increased in the Bay watershed from 611,017 to 860,004 ac between 1990 and 2000. These impervious surfaces collect runoff and prevent it from infiltrating into soils and surface sediment, where rainfall can be stored and natural filtering can often remove some pollutants. These paved areas also contribute significant pollution (e.g., oils, pesticides, herbicides, and nutrients) to waterways.

QUESTION 10-13. Calculate the annual rate of increase of impervious surface from 1990 to 2000. Use $r = (N - N_0)/N_0$ or $r = (1/t)\ln(N/N_0)$ (see pp 94–95 and Appendix 1).

QUESTION 10-14. How does the annual rate of increase of impervious surface compare with the annual rate of increase of population? Discuss your answer.

Approximately 100 acres of forest are lost daily in the Bay's watershed. The current forested area of 24 million acres represents 58% of the Bay's watershed. In the 1600s, about 95% of the watershed was forested.

QUESTION 10-15. Calculate the annual rate of forest loss, starting with a forested area of 24 million acres (note that the rate of loss above is daily).

■ A New Bay Agreement

On June 28, 2000, the Chesapeake Bay Program adopted a new Bay agreement, Chesapeake 2000: A Watershed Partnership. The Chesapeake 2000 agreement directs the Bay Program remediation efforts from 2000 to 2010.

In 2003, the Chesapeake Bay Program published a status report on remediation efforts. Here is a summary of that report.[4]

- Bay grasses covered over 68,000 acres of bottom, about 10% of their original distribution. This represented a doubling of area since 1991.

- Almost 500 miles of riparian forest had been restored between 1996 and the end of 1999. (The goal of 2000 miles by 2010 has already been met.) Eighty thousand miles of stream bank would have to be restored to return the watershed to its pre-European condition.

- The nitrogen load declined 42 million pounds per year between 1985 and 1998.

- The phosphorus load declined 6 million pounds per year.

- The number of boat waste pump-out facilities increased to nearly 600 by the end of 1999 in Virginia and Maryland. Human waste discharged from boats is an important source of nitrogen and phosphorus to the Bay.

- The Bernie Fowler "sneaker index" rose to nearly four feet. Bernie Fowler was a Maryland legislator revered for his work toward Bay restoration. Each year Fowler waded into the Bay at the same place and measured the depth at which he could still see his sneakers. In the 1950s, he could see them in more than 5 feet of water, but by 1989, the sneaker index was below 1 foot! Water clarity is one important indicator of water quality.

- Between 1988 and 1999, 1,032 miles of tributary streams were reopened to migratory fish passage. There were more than 2000 dams and impoundments on the Bay as of the early 1980s. These dams block migratory fish from reaching their historic spawning grounds.

---

4. Available at www.chesapeakebay.net/status_indicatorsbackground.aspx?menuitem=28514.

Let's see how successful selected restoration efforts have been. Table I 10-1 contains estimates from aerial surveys of the area covered by sea grasses. Figure I 10-4 shows graphically how successful pollution reduction actions have been as of 2007.

QUESTION 10-16. Convert the area covered by seagrasses from ha to ac (1 ha = 2.471 ac) in Table I 10-1. Then, on the axes below, or on our website (http://www.jbpub.com/abel/), plot the area in ac by year.

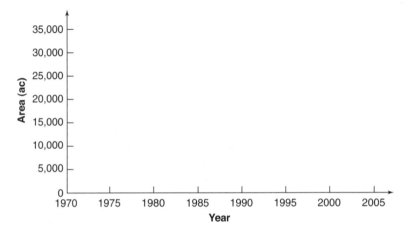

QUESTION 10-17. Based on Table 10-1 only, what could you conclude about the health of sea grasses in Chesapeake Bay from 1970 to 2005?

QUESTION 10-18. The bay-wide restoration goal for seagrasses is 185,000 acresac. By 2007, what percentage of this goal had been met?

QUESTION 10-19. Examine Figures I 10-4 and I 10-5, and summarize the effectiveness of actions taken to reduce pollution in Chesapeake Bay.

TABLE I 10-1	Area covered by sea grasses in Chesapeake Bay by year, 1970–2005 (Source: http://www.vims.edu/bio/sav/Segment AreaTable.htm)	
Year	**Area (HA)**	**Area (AC)**
1970	4,928	
1974	3,296	
1980	6,449	
1985	19,873	
1990	24,292	
1995	24,252	
2000	27,986	
2005	31,671	

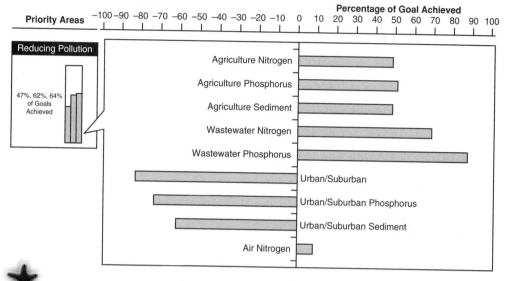

FIGURE I 10-4 Success of actions taken to reduce pollution in Chesapeake bay as of 2007 (Chesapeake Bay Program). Study the figure and describe the actions that have achieved the most success, then list the sources of pollution that most need to be addressed.

Future Challenges

The Chesapeake Bay Program identified numerous challenges to Bay protection that must be addressed in the decades ahead:

- Invasive species, brought into the Bay in the ballast water of ocean-going cargo vessels (see Issue 22)
- Recovery of the Bay's oyster populations, which had virtually disappeared by 1991
- Protection of forest and agricultural land from sprawl development
- Further removal of N and P from sewage
- Opening thousands of miles of tributaries to fish passage by breaching dams or building fish ladders
- Protecting the blue crab, a symbol of Chesapeake Bay but threatened by overharvesting and pollution
- Dealing with material dredged from harbors (Baltimore especially) and shipping channels, which may be contaminated with toxic materials but which, even if uncontaminated, could smother bottom-dwelling organisms

QUESTION 10-20. In light of what you have learned about the state of the Chesapeake Bay, explain whether the population growth rate you calculated earlier sounds large or small.

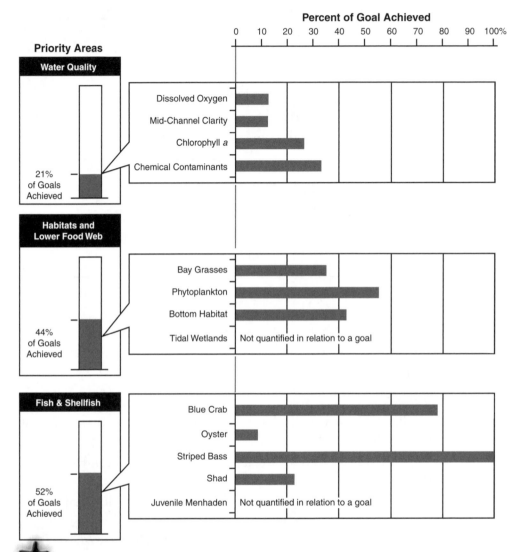

FIGURE I 10-5 Indicators of the health of Chesapeake Bay (Chesapeake Bay Program). How did the Chesapeake Bay Program describe "Bay Health" as of 2007?

Media Analysis 1

Go to www.npr.org, and search for "Chicken Industry Clogging Chesapeake."[5] Listen to the 4-minute program, and then answer the following questions.

5. Or go directly to http://www.npr.org/templates/story/story.php?storyId=18854842.

QUESTION 10-21. Where do most of the nutrients clogging Chesapeake Bay come from?

QUESTION 10-22. To what use is most of the grain grown with the fertilizers put?

QUESTION 10-23. What solution did the scientist Staver offer to the problem of nutrient pollution from manure spread on fields near the chicken operations?

QUESTION 10-24. What results as of early 2008 were described in the report to address the manure problem?

QUESTION 10-25. What was the reason for the suit filed by the Waterkeepers Alliance against the State of Maryland?

Media Analysis 2

Go to www.npr.org, and search for "Lawmakers Create A Stink Over Farm Pollution."[6] Listen to the 4:11 minute program from 2007, and then answer the following questions.

QUESTION 10-26. What substances in manure do environmentalists cite when they contend that manure should be treated as industrial waste?

QUESTION 10-27. What was the pollutant contaminating "Dead Creek"?

QUESTION 10-28. What percentage of the phosphorus pollution in Vermont's Mississquoi Bay was from dairy operations?

QUESTION 10-29. Comment on the statement that farmers forced to pay to remove their pollution would go out of business. What does that suggest about the relationship between the actual cost of their produce and the price it sells for?

QUESTION 10-30. Do you conclude that farms emitting harmful levels of pollution should be held accountable for their action? Why or why not?

QUESTION 10-31. Is "cheap" food an acceptable substitute for a clean environment? Why or why not?

Media Analysis 3

Liverpool's river, the Mersey, was made famous in this country by British rock group Gary and the Pacemakers' song "Ferry 'Cross the Mersey." It is also, of course, the home of the Beatles. The Mersey began to experience serious pollution with the onset of the

---

6. Or go directly to http://www.npr.org/templates/story/story.php?storyId=16126292.

Industrial Revolution in the 1700s. By the 1970s, it had the dubious distinction of being Britain's most polluted river. In 1985, an effort called the Mersey Basin Campaign was begun to attempt to reclaim the estuary. To learn about the Mersey, its sources and types of pollution, and the progress made to clean the river, go to http://www.merseybasin.org. uk/information.asp. Click on "River Mersey 6-minute expert," and answer the following questions.

QUESTION 10-32. What is the human population of the Mersey Estuary region?

QUESTION 10-33. When did the Mersey begin to experience serious pollution?

QUESTION 10-34. What industrial activities have contributed pollution to the Mersey?

QUESTION 10-35. List the types of pollutants dumped in the Mersey mentioned in the report.

QUESTION 10-36. What is the major source of hypoxia (lack of dissolved oxygen) in the Mersey?

QUESTION 10-37. What installations at Sandon Dock helped to restore dissolved oxygen?

QUESTION 10-38. Explain the effect of installing tertiary sewage treatment on the Mersey's environment.

QUESTION 10-39. As a result of water quality improvements, what fish species and other animals have been reestablished in the waterway?

QUESTION 10-40. How has water quality improved in the estuary between 1985 and 2001 from the "Mersey Basin Water Quality" figure on page 4 of the Web report?

ISSUE 11

Dead Zones

What is environmental hypoxia?

What is environmental anoxia?

What causes aquatic hypoxia and anoxia?

What is a dead zone?

How extensive is the problem of oceanic dead zones?

How can the problem of dead zones be addressed?

KEY QUESTIONS

Humankind is engaged in a gigantic, global experiment as a result of the inefficient and often overuse of fertilizers, the discharge of untreated sewage, and rising emissions from vehicles and factories.

—Klaus Toepfer, in a statement accompanying the U.N. Environment Program (UNEP) report *Global Environment Outlook Year Book*

....

I'm convinced this is going to be the biggest environmental issue in the aquatic marine realm in the 21st century.

—Robert Diaz, marine biologist and professor at the Virginia Institute of Marine Science

Hypoxia means dangerously low oxygen, and usually refers to oxygen levels below 2 mg/L. Anoxia refers to water with no dissolved oxygen. Figure I 11-1 shows how an algal bloom can convert waters of the ocean bottom to a hypoxic dead zone. These conditions are replicated each summer in the Gulf of Mexico over an area up to 8,500 square miles (about the area of New Jersey) within about 30 m of the bottom. Any aerobic (oxygen using) organism in this area must either leave or die.

Hypoxia in the Gulf is caused by eutrophication, the result of too many nutrients carried into the Gulf by (1) rivers, primarily the Mississippi and (2) the wind. Although nutrients are essential for the growth of marine plants like algae, too many nutrients can lead to a massive algal growth called an "algal bloom," which can deplete water in oxygen (see Issue 16). Hypoxic conditions can result when the algae die, fall to the bottom, and are decomposed by oxygen-consuming microbes, as shown in Figure I 11-1. Figure I 11-2 shows the extent of the "dead zone" in the Gulf, off the mouth of the Mississippi.

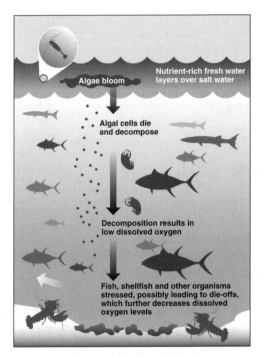

FIGURE I 11-1 The mechanism by which nutrient pollution ultimately can lead to environmental hypoxia or anoxia. Explain in a sentence or two how excess nutrients can lead to hypoxia or anoxia.

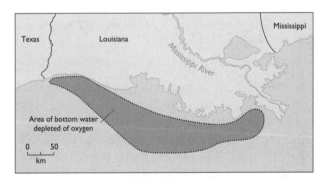

FIGURE I 11-2 A Dead Zone in the Gulf of Mexico off of Louisiana. Based on the scale provided, do a rough calculation of the area of bottom affected. [*Source:* Adapted from U.S. Congress, Office of Technology Assessment. *Wastes in Marine Environments*, OTA-O-334. U.S. Government Printing Office, 1987.]

Where Do Dead Zones Form?

The marked increase in population in coastal regions worldwide (see Issue 2) has contributed to greater nutrient and organic matter inputs into coastal watersheds. This is a main cause of an increase in the frequency and extent of anoxic and hypoxic conditions. Over the last several decades, anoxia has increased in the Baltic Sea, the Black Sea, Chesapeake Bay, and other enclosed or coastal water bodies such as the Gulf Coast. Episodes of hypoxia have also become common in Long Island Sound, the North Sea, and the Kattegat (the strait between Denmark and Scandinavia), resulting in significant declines of commercial fisheries.

Why Does a Dead Zone Form off Louisiana?

According to the Environmental Protection Agency, most of the nutrients entering the Gulf from the Mississippi River arise from human activities. These include discharges from sewage treatment plants, home septic systems, and storm-water runoff from urban areas. Agricultural runoff, from fertilizer and animal excrement, also contributes large amounts of nutrients, especially phosphorus. Storm-water and agricultural runoff are examples of "nonpoint" runoff, that is, runoff not originating from a pipe or other discrete source. Nutrients from automobile exhaust and fossil fuel power plants enter the waterways and the Gulf through air deposition in the Mississippi River watershed. Approximately 90% of the nitrate load to the Gulf comes from nonpoint sources, and more than half of the nitrate load enters the Mississippi River above its junction with the Ohio River. The Ohio River Basin adds 34%.

High nitrogen loads come from wastewater discharges and agricultural lands in Iowa, Illinois, Indiana, southern Minnesota, and Ohio.

Media Analysis 1

Go to www.npr.org, and search for the 2004 report "New Dead Zone Forms Along Oregon's Coast."[1] Listen to the 7-minute report, and answer the following questions.

QUESTION 11-1. Near what coastal city was the dead zone located?

QUESTION 11-2. Describe the physical characteristics of the "dead zone."

QUESTION 11-3. How does the nutrient-rich deep water get to the surface?

QUESTION 11-4. What do oceanographers suggest the origin of the dead zone to be?

---

1. Or go directly to http://www.npr.org/templates/story/story.php?storyId=3850670.

QUESTION 11-5. Describe the oxygen content of the 200-foot deep water sample obtained by oceanographer Francis Chan.

QUESTION 11-6. How may the dead zone be linked to human-caused global warming? Is this a hypothesis, a fact, or a theory? Why?

QUESTION 11-7. Who should be responsible for addressing the dead zone problem? Justify your answer.

QUESTION 11-8. Should fishers be compensated for any financial losses resulting from the dead zone? Explain your answer.

Media Analysis 2

Go to www.npr.org, and search for the 2006 report "The Chesapeake Bay, Scenic and Unhealthy."[2] Listen to the 4-minute report, and answer the following questions.

QUESTION 11-9. What is the nation's largest estuary?

QUESTION 11-10. What evidence did the commenter cite to support his contention that the Bay is "dying"?

QUESTION 11-11. How much would upgrading the Bay's sewage plants and reducing agricultural runoff cost compared with the estimated value of the Bay to the U.S. economy?

QUESTION 11-12. How many new residents settle around the Bay's shorelines yearly?

QUESTION 11-13. What was the number of the Chesapeake Bay Foundation's "Index of Bay Health" in 2006?

Media Analysis 3

Go to www.npr.org, and search for the 2008 report "Low-Oxygen Zones Spreading to Deep Ocean."[3] Listen to the 4-minute report, and answer the following questions.

QUESTION 11-14. What did oceanographer Greg Johnson report had happened to ocean dead zones over the past 50 years?

QUESTION 11-15. How did biologist William Gilley describe the problem of growing anoxia?

QUESTION 11-16. What effect would iron fertilization of the oceans, proposed to remove CO_2 from the atmosphere, have on anoxia?

---

2. Or go directly to http://www.npr.org/templates/story/story.php?storyId=5341055.
3. Or go directly to http://www.npr.org/templates/story/story.php?storyId=90111754.

Ecology of Large Marine Vertebrates

Highly Migratory Fishes

KEY QUESTIONS

What fishes are highly migratory?

What is the importance of highly migratory fishes to marine ecosystems?

What events are responsible for the plight of highly migratory species?

How are bluefin tuna populations threatened?

Populations of many sharks and billfish (marlin, swordfish, etc.), along with other species of big marine fish, are in decline. In 2003, fishery scientists Ransom A. Myers and Boris Worm, writing in the journal *Nature*,[1] concluded that populations of large predatory oceanic fishes, like sharks and tunas (Figure I 12-1), and also demersal (bottom-dwelling) species like cod and flatfish, known to commercial fishers as groundfish, had declined precipitously compared with levels of 50 to 100 years ago.

In a statement released on publication of the article,[2] author Myers said this:

... industrial fishing has scoured the global ocean. There is no blue frontier left. Since 1950, with the onset of industrialized fisheries, we have rapidly reduced the resource base to less than 10%—not just in some areas, not just for some stocks, but for entire communities of these large fish species from the tropics to the poles.

Linda Chandler, of the International Coalition of Fisheries Associations, countered this: "Research shows fisheries are more productive when fished." Moreover, she asserted

---

1. Myers R. A., and B. Worm. 2003. Rapid worldwide depletion of predatory fish communities. Nature 423: 280–283.
2. As quoted in Llanos, Miguel. Study: Big ocean fish nearly gone: Scientists use fishing data to estimate that just 10 percent left. Available at www.msnbc.msn.com/id/3339910/.

FIGURE I 12-1 Northern bluefin tuna (*Thunnus thynnus*).

that fish populations respond to new predators like humans "by reproducing more," as long as the predation is not carried too far.

Who is right, Myers or Chandler? Are "entire communities" of these fish in peril? Or are marine fisheries a necessary component of keeping fish populations healthy? We study sustainable fisheries again in Issue 18. In this issue, we analyze the status of populations and the ecological importance of sharks, tunas, and billfish which are collectively known as *highly migratory species*.

QUESTION 12-1. In a more recent article,[3] Myers and Worms state that "it has long been assumed that these species are largely extinction-proof." Why do you think that many have assumed that large oceanic fishes like tunas and sharks are extinction proof?

Highly Migratory Species

Highly migratory species are oceanic fishes and cetaceans that are widely distributed and migrate large distances to feed or reproduce. They include species popular in oceanic commercial and recreational fishing (e.g., bluefin, albacore, skipjack, and yellowfin tuna,

---

3. Myers R. A., and B. Worm. 2005. Extinction, survival, or recovery of large predatory fishes. *Philosophical Transactions of the Royal Society B* 360:13–20.

and marlin, sailfish, and swordfish, and some sharks), as well as species that are not targeted by fishers. This latter group, known as bycatch, bykill, or simply "trash fish" (see Issue 19), consists primarily of sharks (among the highly migratory fishes) captured on longlines or in drift nets.

There is debate among the fishery community—scientists and fishers—concerning the accuracy of Myers and Worm's conclusion about the magnitude of the decline of the big fishes, but there is scientific consensus that large oceanic fishes are not nearly as abundant as they once were. This determination led the U.S. National Marine Fishery Service to issue the Consolidated Atlantic Highly Migratory Species Fishery Management Plan, which specifies ways to promote recovery of migratory fish species whose populations are depleted. In this Issue, we focus on two of the species managed by the plan, bluefin tuna and sandbar sharks.

■ Bluefin Tuna

The common name bluefin applies to several species of tuna, but most commonly to the Northern bluefin tuna, *Thunnus thynnus* (Figure I 12-1). This species is widely distributed in subtropical and temperate waters throughout the Atlantic and Pacific Oceans. The largest of the tunas, over its 15 to 20+ year life span bluefins can grow to a length of 180 inches (460.0 cm) and over 1,500 pounds (680 kg). At sexual maturity (5 to 8 years), bluefins annually produce about 10 million eggs. In Atlantic populations, spawning apparently occurs in the Mediterranean and Gulf of Mexico. It occurs off the Philippines in the Pacific Ocean.

Bluefins are known for their seasonal migrations. Studies of tagged fish have revealed migrations across both the Pacific and Atlantic Oceans, the latter in less than 60 days!

Physiologically, bluefins and other tunas, along with a few sharks like the mako and great white, have the ability to keep their body temperature higher than that of the water in which they swim (see Chapter 7). This adaptation, which is very rare in fishes because it is difficult to prevent heat loss to water, is known as *endothermy*. The main advantages of endothermy are that a warmer body translates into more muscle power and faster swimming and in some fishes better sensory abilities, all of which makes these fishes more efficient predators. Endothermic fishes can also migrate through water of varying temperatures more readily than other fishes. The cost of being endothermic is that producing so much heat requires lots of fuel, and thus, endothermic fish like the bluefin eat a lot.

Bluefin tuna are a popular food and game fish. In Japan, where they are prized for sushi, individual fish have sold for more than $100,000 U.S. at Tokyo's Tsukiji fish market (Figure I 12-2).

The Consolidated Atlantic Highly Migratory Species Fishery Management Plan considers bluefin tuna stocks as overfished.

QUESTION 12-2. Review the biology and life history characteristics of the bluefin tuna. Identify features that would tend to protect the species from being overfished. Which might tend to hinder recovery from overfishing? In what ways?

FIGURE I 12-2 Bluefin tuna at Tokyo's Tsukiji fish market.

Atlantic Bluefin Tuna in Northern Europe

We tend to think that collapse of marine fisheries is a recent phenomenon because of the combination of improved technology and higher demand for fish. However, there are numerous cases of fisheries collapsing in the early to mid 20th century, and

TABLE I 12-1	Catch rate (in tonnes) for bluefin tuna from 1927–1950 for six European countries. (Source: MacKenzie, B.R. and R.A. Myers. 2007. The development of the northern European fishery for north Atlantic bluefin tuna *Thunnus thynnus* during 1900–1950. Fish. Res. 87: 229–239.)

Year	Denmark	France	Germany	The Netherlands	Norway	Sweden
1927		7.6			50	
1928		51.2	30		116	
1929		91			131	
1930		47			61	
1931		98	33		59	
1932	118	48	54		83	
1933	36	39	26		44	7
1934	21	12	22		59	14
1935	44	23	21		152	
1936	165	2	20		98	9
1937	165	6	14		114	17
1938	180	6	123		177	365
1939	139	0	88		139	527
1940	600	0	0		133	681
1941	108	0	0		269	436
1942	810	0	0		455	2068
1943	80	0	0		72	4
1944	380	0	0		377	221
1945	550	0	0		722	542
1946	590	0	0		227	179
1947	392	0	1	4	210	298
1948	475	0	7	18	368	127
1949	2031	158	169	9	2562	556
1950	991	62	230	10	1712	94

one of these is the European fishery for Atlantic bluefin tuna. A Mediterranean fishery for bluefin tuna dates back to ancient Rome. From the 1940s to the 1960s, there was a major fishery for bluefin tuna in the Norwegian Sea, North Sea, and nearby areas. Table I 12-1 shows catch rate (in tonnes) for bluefin tuna from 1927–1950 for six European countries.

QUESTION 12-3. On the axes on the next page, or on our website (http://www .jbpub.com/abel/), plot the Danish, Norwegian, and Swedish landings of bluefin tuna from 1927–1950.

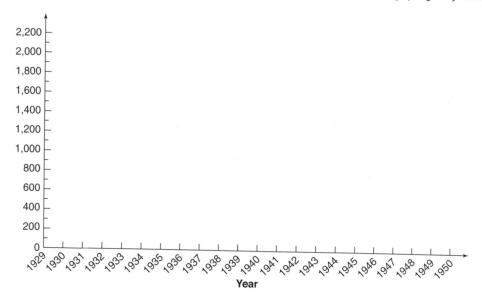

QUESTION 12-4. Interpret your graph. (What happened to catch rate in France and Germany from 1939–45 (i.e., the war years)?

The situation is even more dire in the Mediterranean, where the combination of poor management and pirate fishing (catching more than the legal quota) may lead to the disappearance of the species. Purse-seiners are especially efficient at catching the species, and a new form of exploitation, tuna ranching, is becoming an even greater threat. In tuna ranching undersized tuna are captured and placed in pens where they are fattened before exportation to Japan and other countries.

A Greenpeace report *Where have all the Tuna gone?*[4] concludes

- The 2005 catch in the East Atlantic and Mediterranean exceeded the quota of 32,000 tonnes by 12,000 tonnes. Most of the overage occurred in the Mediterranean.
- The catch included large quantities of immature tuna.
- The problem is complicated because several countries hide or falsify catch data.
- The fishery is highly subsidized, as much as $34 million since 1997 in the European Union.

"My big fear is that it may be too late," said Sergi Tudela, a Spanish marine biologist with the World Wildlife Fund. "I have a very graphic image in my mind. It is of the migration of so many buffalo in the American West in the early 19th century. It was the same with bluefin tuna in the Mediterranean, a migration of a massive number of animals. And now we are witnessing the same phenomenon happening to giant bluefin tuna that we saw happen with America's buffalo. We are witnessing this, right now, right before our eyes."[5]

---

4. Available at http://oceans.greenpeace.org/raw/content/en/documents-reports/tuna-gone.pdf.
5. Quoted in http://ngm.nationalgeographic.com/ngm/0704/feature1/.

QUESTION 12-5. Suggest and evaluate ways to increase and stabilize stocks of Atlantic bluefin tuna in Northern Europe.

· ·

Media Analysis

Go to www.npr.org, and search for "Tracking Bluefin Tuna."[6] Listen to the 5-minute report and answer the following question.

QUESTION 12-6. Explain how the Stanford University biologist's findings may change the way bluefin tuna are managed.

---

6. Or go directly to http://www.npr.org/programs/morning/features/2001/aug/bluefintuna/010817.bluefintuna.html.

The Demise of Sharks

Are sharks populations threatened?

What roles do sharks play in their environment?

Could declines in shark populations lead to extinction?

Would the consequences of shark extinctions be entirely negative, especially because sharks consume economically important fish and they also periodically "terrorize" swimmers?

What measures can be taken to help populations of highly migratory fishes, especially sharks, recover?

KEY QUESTIONS

He was a very big Mako shark, built to swim as fast as the fastest fish in the sea and everything about him was beautiful except his jaws. His back was as blue as a swordfish's and his belly was silver and his hide was smooth and handsome. He was built as a swordfish except for his huge jaws, which were tight shut now as he swam fast, just under the surface with his high dorsal fin knifing through the water without wavering.... This was a fish built to feed on all the fishes in the sea, that were so fast and strong and well armed that they had no other enemy.

—Ernest Hemingway, from the *Old Man and the Sea*

It is difficult to avoid sharks, at least in the media. A picture of a leaping great white shark about to capture a commando dangling from a helicopter, actually a composite hoax, spread quickly over the Internet. Sharks played prominent roles in the movies *Finding Nemo*, *Shark Tale*, and *Open Water*. Fishing for sharks is still popular (Figure I 13-1), and shark attacks, however rare, still command media attention.

Based on all of this attention, one might be led to conclude that sharks are prospering. In fact, the opposite is true. The IUCN (World Conservation Union) lists 66 species of sharks and their close relatives, rays, as threatened, endangered, or critically

FIGURE I 13-1 A boy poses with a male Atlantic sharpnose shark, a common coastal species, before releasing it. Catch-and-release fishing is becoming popular among recreational shark fishers, but unanswered questions remain about post-release mortality. [Photo by D. C. Abel.]

endangered. Rapid declines in populations of large coastal and oceanic sharks have occurred in the last 15 years.

Thus, after dominating the oceans for most of their nearly 400 million year existence, the long-term survival of many kinds of sharks and rays is in question. By 1999, the number of sharks along the East coast of the United States had declined so precipitously that the National Marine Fisheries Service, in accordance with the Magnuson-Stevens Fishery Conservation and Management Act, implemented a revised Fishery Management Plan to rebuild stocks of over 70 species of sharks (as well as other highly migratory fishes like swordfish and tuna).

Other countries and international organizations have also taken actions to protect sharks. The Department of Fisheries and Oceans of Canada issued an Atlantic Shark

Management Plan in 1997. In 1999, member countries of the Food and Agricultural Organization of the United Nations endorsed a shark management agreement known as International Plan of Action for Sharks. Also in 1999, the U.S. House of Representatives unanimously endorsed a ban on the destructive practice of shark finning after it came to light that 55,000 blue sharks were killed in 1998 for their fins, which are the key ingredient in shark-fin soup, an Asian delicacy. In 2004, more than 60 countries agreed to prohibit the killing of sharks for their fins in the Atlantic Ocean, and the Convention on International Trade in Endangered Species listed three sharks (great white shark, basking shark, and whale shark) to receive international protection through trade restrictions.

Ecology of Sharks

In coastal and oceanic ecosystems, carnivorous sharks occupy a position at the top of food webs and chains as apex predators (see Chapter 7). Within an ecosystem, apex predators exert a strong influence on the organisms below them in the food web. First, they may play a major role in controlling both the diversity and abundance of other species within a community. Second, they likely influence the evolution of other species. In both cases, they do this through predation on sick, injured, slow, weak, or otherwise less fit individuals.

Threats to the Survival of Sharks

For any species or population of organisms, there exists a population size below which that group is doomed to extinction. This is known as a *critical number*. When the number of adult individuals drops to this level, there is simply an insufficient reservoir of genetic variability and potential for mating to allow the population or species to survive. Genetic variation is central to the survival of a species because it is the raw material for natural selection. When a population's environment changes, genetic variation can produce some individuals that have the characteristics necessary for survival.

Critical numbers, however, are difficult to determine, and efforts to do so are often not undertaken until a species is near extinction.

Several species of sharks and rays may be threatened with extinction because of overfishing and habitat alteration. Annual catches of sharks, skates, and rays had reached 800,000 metric tons (nearly 2 billion pounds) by 2000. Fishing pressure is likely to increase because most shark fisheries are still small scale. As a transition to more modern, industrial-type fisheries has occurred, landings have increased substantially. Sharks are also frequently captured on longlines and in nets as bycatch, that is, as untargeted and hence unused catch (see Issue 20).

Sharks are particularly vulnerable to pressures that can reduce their population because of their life history characteristics.

- There are typically fewer individuals of sharks than other species in marine ecosystems.
- Shark are very slow growing.

- Individuals are long lived and reach sexual maturity late in life, in some cases requiring 15+ years to become mature.
- There is a long gestation period (the time developing embryos are retained in the female or in the egg).
- Sharks typically rely on specific mating and nursery areas.
- Fecundity (the number of offspring produced by a female of a species) is low.

Sandbar Sharks

Let us focus on the sandbar shark (Figure I 13-2), *Carcharhinus plumbeus*, a large coastal species found worldwide and at one time abundant from Cape Cod to Brazil in the western Atlantic Ocean.

Before fishery scientists can determine whether a species is threatened by over-fishing or other impact, it is critical to understand its life history, which is fairly well known for only a handful of the 400+ species of sharks. Estimates of age and growth parameters for sandbar sharks have been obtained from tagging studies and research on captive animals. These results show that sandbar sharks are born at a length of about 45 to 50 cm (16 to 20 in) and grow about 8 to 10 cm (3 to 4 in) the first year. They reach sexual maturity at about 136 cm (53.5 in) at 15 years of age and have a life span of about 30 to 35 years (Table I 13-1). Eight to 10 pups are

FIGURE I 13-2 Sandbar sharks caught on a 500-ft long, 25-hook research longline. These animals were measured, tagged, and released relatively unharmed as part of a study by co-author D. Abel at Coastal Carolina University. A recent assessment of this species in the Atlantic found that overfishing is occurring. [Photo by D. C. Abel.]

born every other or every third year after a 12-month gestation period. In contrast, a typical bony fish such as a trout or cod reaches maturity earlier, has more young, grows faster, and has a considerably larger population size.

QUESTION 13-1. Explain how these life history characteristics make sharks particularly vulnerable to threats like overfishing and habitat loss.

QUESTION 13-2. Although many life history characteristics of sharks make them vulnerable to overfishing and other threats, there is one advantage in having few offspring and a long gestation period. What do you think this advantage might be? (HINT: Newborn sharks require no parental care).

The most complete data on population levels of sandbar sharks, the most common shark in Chesapeake Bay, are from long-term studies of the Virginia Institute of Marine Science Shark Ecology program. This program sampled eight or more stations in and

TABLE I 13-1   **Life history traits of selected long-lived animals**

Species	Age to Maturity (years)	Life Span (years)	Litter Size	Annual Rate of Population Increase
Sandbar shark (*Carcharhinus plumbeus*)	13–16	35	8–13 pups	2.5–11.9%
Southern bluefin tuna (*Thunnus macocyii*)	10–11	40	14–15 million eggs	?
Atlantic cod (New England stock) (*Gadus morhua*)	2–4	20+	2–11 million eggs	?
Summer flounder (*Paralichthyes dentatus*)	1	F: 20 M: 7	0.5–5 million eggs	50%
Bottlenose dolphin (*Tursiops truncates*)	M: 11 F: 12	50+	1	?
Loggerhead sea turtle (*Caretta caretta*)	12–30	50+	116 eggs/clutch 76.5/year	2–6%
Asian tiger (*Panthera tigris*)	3–7	26	1–7 per litter	?
African elephant (*Loxodonta africana*)	8–13	55–60	1	4–7%

*Modified from Camhi, M. *et al.* 1998. Sharks and their Relatives Ecology and Conservation. Occasional Paper of the IUCN Species Survival Commission No. 20 International Union for Conservation of Nature and Natural Resources. Available at http://www.flmnh.ufl.edu/fish/organizations/ssg/shark2.pdf.

offshore of Chesapeake Bay from May through September or October annually over the period 1980 to the present. Sharks were captured with 100-hook longlines stretched for 1 to 1.5 nautical miles (1.8 to 2.8 km) and fished 3 to 4 hours. The longline configuration represents what fisheries scientists call one unit of effort. Longlines are horizontally stretched ropes or heavy monofilament with branches containing hooks at fixed intervals. Although this method sounds dangerous, it generally does not harm the sharks (although some species fare more poorly than others), which are simply tagged and released (Figure I 13-2). Commercial longlines, as discussed later here, are a different matter.

Data are reported as Catch Per Unit Effort (CPUE), that is, the number of sharks caught per 100 hooks per 3- to 4-hour fishing time. For example, a CPUE of 2.3 means that, on average, 2.3 sharks were caught per hundred hooks that soaked for 3 to 4 hours. Table I 13-2 shows the mean CPUE for adult sandbar sharks from 1980 to 2007.

QUESTION 13-3. Plot the data from Table I 13-2 on the axes below, or on our website (http://www.jbpub.com/abel/).

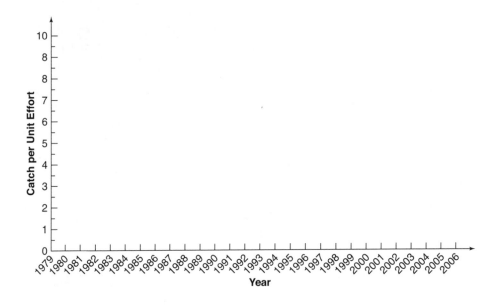

QUESTION 13-4. Interpret the graph. What trend(s) do you infer?

QUESTION 13-5. Think about what could cause a shark population to decline. Then list as many possible explanations as you can for the decline in the sandbar shark population.

QUESTION 13-6. The data on which this analysis was based were obtained by scientists and are known as "fishery independent." Explain whether you think

TABLE I 13-2	Catch Per Unit Effort (CPUE) data for Sandbar Sharks in the Chesapeake Bay Region, 1979–2006. CPUE units are sharks per 100 hooks

Year	CPUE (Sharks per 100 hooks)
1979	9.1
1980	7.1
1981	8.8
1982–1985	2.6
1986–1989	3.2
1990	0.9
1991	1.0
1992	0.3
1993	1.5
1995	2.2
1996	2.7
1997	2.7
1998	3.8
1999	2.9
2000	3.5
2001	3.1
2002	1.8
2003	1.4
2004	2.6
2005	1.8
2006	2.9

data collected this way are better, worse, or no different than data collected and reported by commercial fishers ("fishery-dependent data").

Sandbar sharks are second only to blue sharks in popularity with recreational fishers. Moreover, because there is a lucrative overseas market for shark fins and the sandbar shark's fins are so large, they are even more valuable. Shark fins are used in Asian shark fin soup. Depending on the species from which the fins were obtained, retail prices for shark fins can be as high as $300 a pound, or about $40 per fin.

Finally, sandbar sharks are a popular food fish along the U.S. East Coast.

QUESTION 13-7. Sharks have become very popular as an upscale food fish in the last 20 years. Suggest reasons for this.

■ Promoting Recovery of Sandbar Sharks

A National Oceanic and Atmospheric Administration study suggested that annual catch of sandbar sharks must be reduced by about 78% in order for the population to recover by the year 2070. Public policy goals, such as helping depleted stocks of sandbar and other sharks to recover, can be achieved through prohibition, direct regulation, moral persuasion, payments and incentives, and litigation (lawsuits). Here are some possible measures or activities that could promote recovery of sandbar sharks.

- A "shark stamp" purchased by recreational fishers, with proceeds going to promote conservation
- Banning shark fishing entirely
- Taxes or impact fees for boats and/or fuel
- Hiring more staff to patrol waters and enforce regulations where shark fishing occurs
- Selecting the sandbar shark as the "official shark" of Virginia, North Carolina, and so forth
- Advertising campaigns on television and in other media
- Mandatory conservation classes for recreational fishers
- Bans on selling shark meat at fish markets
- Incentives to catch fish other than sharks
- Limitations on the number of shark fishing permits
- Grants for teaching shark conservation to public school students
- Banning the sale or import of shark jaws, teeth, and so forth
- Filing lawsuits against violators of current shark fishing restrictions
- International trade agreements restricting sale of shark fins
- Government buy-backs of shark-fishing boats
- No-take zones

QUESTION 13-8. Select which methods from the above list you think would be most effective. Identify the drawbacks?

QUESTION 13-9. What problems might you face with trying to manage a highly migratory species?

Table I 13-1 shows selected life history traits of some long-lived animals.
QUESTION 13-10. Examine Table I 13-1 and identify which animals have life history characteristics most like that of the sandbar shark (the first species in the table).

QUESTION 13-11. In light of your answer to the previous question, do you think that sharks whose populations are threatened can be managed like the other fish species listed in Table I 13-1? Explain your reasoning.

The Consequences of the Loss of Apex Predators

What would the consequences be if top predators like sharks, typically the least abundant organisms in an ecosystem, disappeared from the ocean? There are several examples from the terrestrial environment—the loss of wolves, bear, and cougars in the western United States led to increases in populations of their prey, elk, deer, coyote, and so forth, which caused ecological changes in plant communities. Such far-reaching effects are known as ecological or trophic cascades. Would a decrease in the number of sharks, or their disappearance, cause a similar marine ecological cascade, an ecological domino effect?

A recent study[1] provides the first evidence that the loss of sharks may indeed cause a trophic cascade. The study was based in part on the University of North Carolina's shark survey, conducted annually since 1972, which showed declines in the following species from 1972 to 2006: sandbar sharks, 87%; blacktip sharks, 93%; tiger sharks, up to 97%; scalloped hammerheads, 98%; and bull, dusky, and smooth hammerhead sharks, 98% or more.

The researchers then examined population data for one major category of the prey of these large sharks—smaller sharks and rays, so-called *mesopredators*. These elasmobranch (shark and ray) mesopredators are large enough that their only predators are bigger sharks. This kind of feeding relationship is known to ecologists as top-down regulation.

The study indeed found large increases in the populations of most of these mesopredators, including Atlantic sharpnose sharks (Figure I 13-1) and cownose rays. The cownose ray population in Chesapeake Bay is thought to exceed 40 million, and they consume large amounts of bivalves, particularly scallops.

According to the study, in 1983 and 1984, when cownose ray populations were considerably smaller, they had no impact on scallop populations. By the late 1990s, these rays caused complete mortality in areas where they fed. In North Carolina, by 2004, the bay scallop fishery crashed, a phenomenon the authors attribute to the impact of the cownose rays.

Julia Baum of Dalhousie University, one of the study's authors, stated, "Large sharks have been functionally eliminated from the East Coast of the U.S., meaning that they can no longer perform their ecosystem role as top predators."[2]

QUESTION 13-12. Should sharks be protected globally? Summarize the benefits and problems associated with global protection of sharks.

QUESTION 13-13. Many people think the oceans would be a safer place if all sharks disappeared. Explain whether you agree or disagree, and support your position with evidence.

1. Myers, R. A., J. K. Baum, T. D. Shepherd, S. P. Powers, C. H. Peterson. 2007. Cascading effects of the loss of apex predatory sharks from a coastal ocean. *Science* 315:1846–1850.
2. Quoted in http://www.somd.com/news/headlines/2007/5253.shtml.

. . . .

Media Analysis 1
. . . .

Go to www.npr.org, and search for the 2003 report "Drop in Shark Numbers Startles Researchers."[3] Listen to this 4-minute report. Then answer the following questions.

QUESTION 13-14. Why do you think scientists have not put much effort into tracking shark populations? Do you think scientists track populations of other animals? Plants? Explain your answer.

QUESTION 13-15. What method did the researchers use to get a glimpse of shark population trends?

QUESTION 13-16. What did the scientists find?

QUESTION 13-17. If the sharks in the study are not fished commercially, how can the population declines be explained?

QUESTION 13-18. Who is Russell Hudson and why does he dispute the contention that shark populations are collapsing?

QUESTION 13-19. Many people like Mr. Hudson earn their livings in ways that are harmful to the environment. How would you deal with people whose livelihoods negatively impact the environment? What weight would you assign to jobs when considering environmental protection?

QUESTION 13-20. Who is Sonja Fordham, and why does she urge fishing regulators to be more cautious in regulating sharks?

QUESTION 13-21. What does researcher Julia Baum mean when she distinguishes between "extinction" and "extirpation"?

QUESTION 13-22. Explain what you think is meant by the statement at the beginning of the report that declines in shark populations could harm the entire food chain.

---

3. Or go directly to http://www.npr.org/templates/story/story.php?storyId=923750.

Lifestyles of the Large and Blubbery: Of Blue Whales and Krill[1]

KEY QUESTIONS

What enables the blue whale to grow so large?

Why is the Antarctic marine ecosystem so productive?

What are krill, and why are they considered a "keystone species"?

What are the threats to the Antarctic marine ecosystem?

The blue whale (*Balaenoptera musculus*) (Figure I 14-1) is the largest animal that ever lived, reaching a length of about 30 meters (98 feet) and weighing around 100 tons (91,000 kg). Have you ever wondered why this is the only animal that has grown to this size? Exactly what factors allow this species to achieve its dimensions but inhibit others from doing so?

Antarctic Marine Ecology

The blue whale spends its summers in the waters surrounding Antarctica. Antarctica is completely surrounded by the southernmost extensions of the Pacific, Atlantic, and Indian Oceans, a physically distinct body of water collectively referred to as the Southern Ocean. The closest continent, South America, is 970 km (600 miles) away.

Antarctica is, of course, a land mass, but you would not know it from looking at it: Nearly 98% of its 14 million km^2 (5.4 million mi^2) is covered with ice up to 4.5 km (nearly 3 miles) thick. Including sea ice, Antarctica holds 90% of the world's ice and 70%

---

1. Steve Berkowitz of the Department of Marine Science of Coastal Carolina University contributed to issue.

FIGURE I 14-1 The blue whale (*Balaenoptera musculus*), the largest animal that ever lived.

of the world's fresh water. Ironically, because there is no rain in the interior (less than 4 cm of fine ice crystals known as "diamond dust" per year), the continent is essentially a barren desert.

The Southern Ocean

In contrast to Antarctica, the Southern Ocean supports a richly productive assemblage of organisms. What makes these cold waters so productive is the southerly flow of deep, nutrient-rich water known as Circumpolar Deep Water. This water rises to the surface (upwells) near the continent from depths of 3,000 meters (9,800 feet) and makes nutrients available to photosynthetic organisms (principally diatoms), which live in sunlit shallow waters and festoon the underside of sea ice. This upwelled water also transfers heat to the Antarctic atmosphere. In addition, at Antarctic latitudes, photosynthesis can occur 24 hours a day during the summer.

The northern limit of this productivity is a narrow belt of water from 20 to 30 miles (32 to 48 km) wide known as the Antarctic Convergence, or Polar Front (Figure I 14-2). Here, cold Antarctic surface waters sink below warmer waters flowing south. This is also the northern boundary of the Southern Ocean.

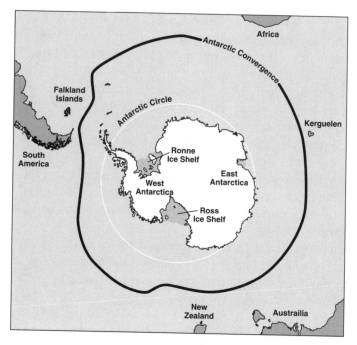

FIGURE I 14-2 The Antarctic Convergence surrounds Antarctica and marks the approximate boundary of the Southern Ocean.

Krill

The key indicator of this Antarctic productivity, in terms of biomass, is not the diatoms or other tiny photosynthetic organisms that actually use the nutrients in the first place. It is a 5- to 6-cm (2 to 2.5 in) long crustacean called Antarctic krill (*Euphausia superba*) (Figure I 14-3), which can top the scales at 2 g (0.1 ounces).

FIGURE I 14-3 Antarctic krill (*Euphausia superba*), a keystone species in the Southern Ocean. Adults may grow to 6 cm and may weigh 2 grams.

Krill, meaning "young fish" in Norwegian, is a term applied to a group of about 85 species of shrimp-like organisms collectively called euphausiids (phylum Arthropoda) that inhabit waters from the poles to the tropics. Krill feed on tiny diatoms in the Antarctic. They are heavier than water and must swim continuously to avoid sinking. Although the life span of krill is not known with certainty, scientific estimates range from 5 to 11 years. Krill are in turn eaten by whales, seals, birds, squid, and fish, and to a lesser (but growing) extent, humans.

Krill might be the most abundant animal species on Earth. Dense aggregates weighing an estimated 2 million tonnes and covering as much as 450 km^2 have been observed. Density of krill schools reportedly reach 30,000 animals per cubic meter of seawater! The total standing stock (the biomass at any time) has been estimated to be between 200 and 700 million tonnes. In contrast, the 2004 total world fish catch was about 95 million tonnes.

Thus, the numerical abundance and biomass of krill, as well as its important position in the Antarctic food web (see Figure I 14-4), make it a keystone species in its ecosystem. A keystone species is one whose impact on its ecosystem is disproportionately large and whose loss would severely disrupt the system.

Krill Fisheries

Before the advent of European whaling in the 19th century, baleen whales consumed huge volumes of krill. After the near-extinction of these whales by the 1960s, some viewed krill to be present in "surplus" amounts (estimated at 150 million tonnes!) which could be harvested without impacting either the species of the marine environment.

Large-scale fisheries for krill currently occur only in Antarctic waters and off the coast of Japan. In the mid 70s, the Soviets were the first to operate full-scale fisheries for Antarctic krill. Japan, Ukraine, and Poland currently conduct Antarctic krill fisheries.

As conventional world fisheries have declined from overfishing, there has been greater emphasis on increasing commercial catches of species from more distant waters. This increase in demand will lead to a greater pressure on Antarctic krill, which are the largest known krill stock.

Currently, most krill taken are used as bait and aquaculture feed, although a small but significant proportion is used for human consumption. Processed krill are fed to farmed fish because of their nutritional value and also because krill contain high concentrations of the red pigment group carotenoids, which heightens the red color of fish. Japanese consumers consider red a sign of good luck and an appetite stimulant.

The Antarctic krill fishery is currently not increasing because of the immense expense of conducting a fishery in Antarctic waters; however, it is likely that this will change because of the growing pressure to catch krill commercially for use as feed in aquaculture. Krill fisheries may also respond to demand for krill concentrate, a "health food" because

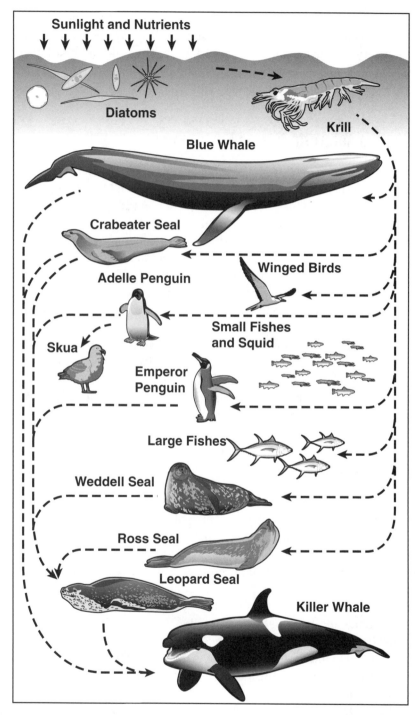

Sunlight and Nutrients

Diatoms

Krill

Blue Whale

Crabeater Seal

Winged Birds

Adelle Penguin

Small Fishes and Squid

Skua

Emperor Penguin

Large Fishes

Weddell Seal

Ross Seal

Leopard Seal

Killer Whale

FIGURE I 14-4 Simplified Antarctic marine food web. Describe the importance of diatoms in this food web. What is the role of krill?

of its omega-3 fatty acids content. What this portends for krill is not known because fishery managers have little experience in dealing with an organism such as krill, which occupy a lower trophic tier in food chains and webs. Thus, although krill fisheries may increase the world's ocean harvest, they also could cause harm, perhaps catastrophic harm, to this species and to marine ecosystems.

Threats to the Antarctic Marine Environment

Even though it is far removed from the kinds of environmental problems that most of the world is facing (fisheries notwithstanding), the Southern Ocean ecosystem is facing some major threats, all of which may have an impact on krill populations.

Temperature increases caused by global warming are accelerating melting of ice sheets and adding massive volumes of freshwater to the marine environment. This could disrupt the sinking of cold, dense water masses (thermohaline circulation or meridional overturning circulation), which is a major source of the world's bottom water. The effects of disrupting the flow of bottom water cannot be predicted, but they would certainly be severe. Additionally, sea ice may provide protection for krill, which as we mentioned live on the bottom of sea ice.

Antarctic marine organisms would be particularly susceptible to temperature increases because they live (and have evolved in) a relatively stable thermal environment. Like all organisms, Antarctic organisms live closer to their thermal maximum (the highest temperature they can tolerate) than their thermal minimum. Thus, increases in temperature could lead to a host of lethal and sublethal effects.

Ecological Energetics

In this analysis, you learn about ecological efficiency and calculate the mass of diatoms it takes to ultimately "build" both an orca and a blue whale.

As previously mentioned, the cold Antarctic waters are extremely productive. This productivity is fueled by the upwelling of nutrient-rich water, as well as the long photoperiod (period of light). It also depends on the number of links in the individual food chains comprising its food web—the fewer links, the more production. Biological processes are far from 100% efficient. Thus, at each step in a food chain, when one organism consumes another, only a small fraction, typically 10% to 20%, of the energy value of the food is incorporated into the body of the grazer or predator. The rest of the energy is consumed in metabolism or is dissipated as heat. Thus, the more steps there are between the photosynthetic organisms at the base of the food web and the top predator, the greater is the energy loss.

Figure I 14-4 depicts a simplified part (the pelagic portion) of an Antarctic food web. The arrows indicate the direction of energy flow. One pathway leads from diatoms to krill to squid to Weddell seals to leopard seals to orcas. This pathway thus has six levels with five intervening steps. A second pathway has three levels and two steps: diatoms to krill to blue whales (which are filter feeders).

This figure shows that krill are eaten by many winged birds and penguins and by fishes, squid, and species of seal. Two species of penguin, the Adelie (which is the most abundant penguin in Antarctica) and the chinstrap, live predominantly on squid. The population sizes of these two species are orders of magnitude greater than the larger emperor and king penguins, whose diet is more diverse.

The most common seal, the crabeater, has a diet based almost exclusively on krill. Its population size, estimated at 15 to 40 million, is greater than that of all the other Antarctic seals combined. The blue whale illustrated in this food web is not the only krill-feeding whale in Antarctic waters. Right, minke, sei, fin, and humpback whales are also found in the Southern Ocean.

Now, let us calculate how many kg of krill would be produced by consumption of 1 kg of diatoms. If we assume an ecological efficiency of 15% (0.15), then we simply multiply 1 kg (the mass of diatoms) by 0.15 (the efficiency of energy transfer), and the answer is .15 kg, or 150 g.

For the following calculations, we assume that the ecological efficiency is 15%.
QUESTION 14-1. How many kilograms of killer whale would be produced by 10,000 kg of diatoms?

QUESTION 14-2. Let us look at this from a different perspective: How many kilograms of diatoms are needed to build 1 metric ton (1,000 kg) of killer whale?

QUESTION 14-3. How many kilograms of diatoms are needed to build 1 metric ton (1,000 kg) of blue whale?

QUESTION 14-4. This diatoms-krill-blue whale pathway is a very efficient and unusually short food chain. It is thought that the process of filter feeding allows this pathway to exist. How much more efficient is this diatoms-krill-blue whales food chain than the longer one we examined in Question 14-1?

QUESTION 14-5. Explain why filter feeding by blue whales enables them to grow so large.

Ozone Depletion

In this analysis, we examine the possible impact of atmospheric ozone depletion on marine phytoplankton and then evaluate the impact on blue whales and orcas.

Around 1985, scientists became aware that ozone levels in the Antarctic stratosphere (upper atmosphere) were dropping significantly each spring. The source of the degradation was a group of synthetic chemicals called chlorofluorocarbons, which were and still are used as propellants and refrigerants. Under the Antarctic atmospheric conditions (extreme cold and sunshine), these chemicals acted as powerful ozone depleters. The result of this ozone depletion has been an increase in ultraviolet radiation reaching the surface of the Earth and penetrating the top meter or so of seawater. High intensities of ultraviolet radiation have the potential to harm unprotected phytoplankton and could decrease phytoplankton production by an order of magnitude or more.

> **QUESTION 14-6.** Phytoplankton production (and population size) can be reduced by the increased intensity of ultraviolet radiation. If the starting diatom biomass from Question 14-1 (10,000 kg) is reduced by 12%, how many kg of killer whale would then be produced (use the same 15% ecological efficiency value)?

> **QUESTION 14-7.** If the starting diatom biomass from Question 14-1 (10,000 kg) is reduced by 12%, how many kg of blue whale would then be produced (use the same 15% ecological efficiency value)?

> **QUESTION 14-8.** Is the potential decrease in blue whale and killer whale mass, resulting from decreased diatom production as a result of ozone depletion, cause for concern? Why or why not?

· ·

Krill Fisheries

Here you examine krill fisheries and speculate on the impact of current and future fisheries on the Antarctic marine ecosystem.

Firms have proposed mechanized harvesting of krill by factory ships from India, Canada, the United Kingdom, the United States, Norway, and Australia. Japanese ships already harvest a few hundred thousand metric tons a year, and before its breakup, the USSR fleet harvested nearly a million metric tons a year during the late 1980s.

> **QUESTION 14-9.** What would be the impact on blue whales should large-scale "mining" of krill take place? On Weddell seals? On orca? On other species? What additional information would you need to answer this question fully?

> **QUESTION 14-10.** Currently, most krill taken in fisheries are used as bait and live-stock feed. In addition, krill are fed to farmed fish to heighten their red color. Comment on the environmental impact of using krill for animal feed.

Krill are threatened by other problems as well. Scientists have observed declines in krill that are linked to diminishing polar ice and warming. In addition, scientists have observed an 8% increase in precipitation in the Southern Ocean over the last 20 years. More increases are expected as a result of climate change due to greenhouse warming. This has the potential to alter the marine ecosystem as well as to disrupt thermohaline circulation.

> **QUESTION 14-11.** Propose solutions to the problems outlined in this issue. What are the impediments to implementing your solutions? Discuss.

QUESTION 14-12. Is one of the lessons from this issue that humans should eat krill or even shrimp (see Issues 18 and 19)? Why or why not?

QUESTION 14-13. Discuss a potential large-scale krill fishery from the perspective of sustainability.

Media Analysis 1

Go to www.npr.org, and search for "Declines Seen in Crucial Penguin Food Staple."[2] Listen to the 3-minute 30-second report, and then answer the following questions.

QUESTION 14-14. Describe the decline in krill in "Krill Central" in the past 30 years.

QUESTION 14-15. Explain why British Antarctic Survey scientists conclude that krill have declined.

QUESTION 14-16. Describe the decline over the past 30 years in Adelie penguins, which depend on krill as a food source.

QUESTION 14-17. Explain whether you can conclude that the decline in krill as a food source is the cause for the decline in the Adelie penguin population.

QUESTION 14-18. Discuss why krill are called a "linchpin of marine life" in the Antarctic at the beginning of this report.

Media Analysis 2

Go to www.npr.org, and search for "Shifting Winds Disrupt Island Birds' Feeding Habits."[3] Listen to the 6-minute 24-second report, and then answer the following questions.

QUESTION 14-19. Why are the Farallons referred to as the "smoke alarms for climate change"?

QUESTION 14-20. How are winds related to the appearance of krill around the Farallon Islands?

QUESTION 14-21. How may global climate change be implicated in the disappearance of key bird species like the Cassin's Auklet?

QUESTION 14-22. How might these changes be related to potential declines in commercially important fish species from Alaska to Baja California?

---

2. Or go directly to: http://www.npr.org/templates/story/story.php?storyId=4142383.
3. Or go directly to: www.npr.org/templates/story/story.php?storyId=11253445.

Oceans Without Cetaceans? The Impact of High-Intensity Sonar

KEY QUESTIONS

What are threats to the survival of cetaceans?

How serious are these threats?

What is the military value of high-intensity sonar?

How do such intense sounds affect marine mammals?

Introduction

You are sitting in your back yard on a sunny day, reading a magazine and basking in the sun. Suddenly and completely without warning, you hear a terrifying, overwhelmingly powerful sound blast that makes you scream in panic and pain and clap your hands over your ears. After what seems an eternity, but lasting only a few seconds, the sound stops. You remove your hands and see blood on your palms, and you hear sound as though your ears, which are experiencing excruciating pain, are filled with water. You rush to call 911 but realize in a panic that you cannot hear a telephone conversation.

What do you do?

Classification and Evolutionary History of Cetacea

Cetaceans are very specialized mammals and include many species of whales and dolphins. Mammals are members of the group called synapsids, vertebrates with a single opening in the side of the skull, which first appeared on earth more than 300 million years ago. Cetaceans belong to the subclass eutheria, warm-blooded animals with many specialized features, including extremely sensitive hearing, a very large brain, and a placenta. Eutheria bear live young after a relatively long gestation period.

Cetaceans evolved from four-legged land-dwelling ancestors sometime before 50 million years ago. Most are predators, but a few whales are giant filter feeders with specialized mouth parts called baleen, which is used to sieve schools of minute crustaceans out of the water. The largest animal on Earth, the blue whale (see Issue 14), is a filter feeder.

Threats to Cetaceans

Cetaceans are at significant risk of extinction unless measures are taken to protect them. The specific threats to their survival include the following:

1. Hunting certain species to extinction, which nearly occurred to many whale groups during the 19th century.

2. Poisoning them with toxic chemicals, which we describe in Issue 8. Nearly all cetaceans contain elevated levels of toxic chemicals in their blubber and milk. The long-term impacts of this are unknown.

3. Causing mass mortalities through the use of deafening, even deadly, high-intensity sounds, mainly from submarines.

4. Destroying their habitat and food supplies through development, overfishing, and the effects of climate change.

In this issue, we consider the mass mortalities of cetaceans that have been linked to military use of high-intensity sounds.

■ Low-Frequency Active Sonar

In June 2000, the U.S. Navy and the National Oceanic and Atmospheric Administration reported the results of a 3-month investigation. The research concluded with near-certainty that beachings and mass mortality of whales on Abaco Island, Bahamas in March of that year were caused by the Navy's use of extremely loud "low-frequency active sonar" (LFA sonar) in the area at the same time. Scientists conducting necropsies on the dead whales found extreme damage to the inner ears of the mammals, which have been known for a century or more to communicate using complex sounds over vast distances in the oceans. LFA sonar emits low-frequency sound at intensities of 215 decibels (dB; see Table I 15-1) at the source, according to the National Marine Fisheries Service.

The Navy received permission in 2002 to deploy two LFA sonar systems, one in the Atlantic and one in the Pacific, for a period of 5 years. These systems were to blanket 80% of the world's oceans with man-made sounds far more intense than levels that are known to disturb cetaceans.

How Sound Is Measured

To understand this issue, you first need to understand the decibel classification system. The "strength" of sound is measured on a logarithmic scale: the decibel scale. Thus, sound at 160 dB is 10 times as intense as sound at 150 dB, and sound at 150 dB is 10 times as powerful as sound at 140 dB, and so on.

Research sponsored by the U.S. Navy concluded that "it is highly unlikely that sound below 160 dB will harm marine mammals." However, a report referenced by the U.S. Navy states that long-term hearing loss in humans is accelerated by chronic exposure over years to sounds at 85 dB or greater in air. For humans, exposure to a sound at intensity of dB 144 for 1 second and 126 dB for 1 minute could cause permanent hearing loss. (Sound intensities at rock concerts can reach 115 dB for minutes at a time.) Estimating hearing impairment in marine mammals is difficult, to say the least. The Navy uses exposure to 180 dB as the level at which 95% of marine mammals could be expected to suffer serious damage and seeks to avoid exposing marine mammals to that level or above. Table I 15-1 shows the dB scale with representative sound levels in air for comparison. Remember that the decibel scale is logarithmic.

QUESTION 15-1. How much more intense is a garbage disposal than a normal conversation?

QUESTION 15-2. How much more intense is the 160 dB level for eardrum rupture than the threshold of pain?

TABLE I 15-1	The decibel scale with reference sounds (www.dangerousdecibels.org)
Source	**Intensity Level**
Threshold of Hearing	0 dB
Rustling Leaves/Breathing	10 dB
Whisper	20 dB
Normal Conversation	60 dB
Busy Street Traffic	70 dB
Vacuum Cleaner/Garbage Disposal (Possible hearing damage after 8 hr)	80 dB
Motorcycle (Hearing damage after 8 hr)	90 dB
Large Orchestra	98 dB
Walkman at Maximum Level/Ambulance Siren (Serious hearing damage after 8 hr)	100 dB
Front Rows of Rock Concert	110 dB
Commercial Jet Takeoff	120 dB
Threshold of Pain	120–130 dB
Military Jet Takeoff	140 dB
Instant Perforation of Eardrum	150–160 dB

QUESTION 15-3. How much more intense is the 160 dB level for eardrum rupture than the level of a motorcycle?

QUESTION 15-4. How much more intense is the 215 dB Sonar pulse at its source than the upper level that would cause eardrum rupture in humans?

■ Whale Beachings and High-Intensity Sound

Whale and dolphin beaching is an oft-observed behavior. Animals swim into shallow water so far as to be trapped on beaches, where they may be asphyxiated, resulting from their inability to expand their lungs. They are often so weak and disoriented that they cannot return to deeper water. How might sound contribute to or cause this behavior? Listen to two media analyses that describe controversy over the Navy's use of various frequencies of high-intensity sonar.

Media Analysis 1

Go to www.npr.org, and search for "Sonar Lawsuit."[1] Listen to the 4-minute report from 2002 on a lawsuit filed by environmental groups against the U.S. Navy. Then answer the following questions.

QUESTION 15-5. What was the event that prompted the audio report?

QUESTION 15-6. Describe the effects "acoustic blasts" have on whales.

QUESTION 15-7. What did the scientific exam (necrosis) on the dead whales disclose?

QUESTION 15-8. What did the examining scientists conclude was the cause of the whales' deaths?

QUESTION 15-9. What was the Navy's response to the lawsuit?

QUESTION 15-10. Describe the controversy over whether the sonar caused the deaths of the whales.

Media Analysis 2

Go to www.npr.org, and search for "Judge: Navy Not Exempt From Sonar Ban."[2] Listen to the 4-minute program from 2008, and answer the following questions.

QUESTION 15-11. How and why does the Navy use high-intensity mid-range sonar?

---

1. Or go directly to http://www.npr.org/templates/story/story.php?storyId=1148004.
2. Or go directly to http://www.npr.org/templates/story/story.php?storyId=18689650.

QUESTION 15-12. What was the argument of the NRDC attorney in support of the lawsuit?

QUESTION 15-13. As a result of the lawsuit, what restrictions were placed on the Navy's use of the sonar?

QUESTION 15-14. What was the NRDC's position on the waiver?

QUESTION 15-15. Is "national security" sufficient grounds to warrant waiving environmental laws? Why or why not?

QUESTION 15-16. Some experts are skeptical of the Navy's assertion that there are "hundreds" of silent submarines plying the world's oceans. Assuming it to be factual, is this sufficient reason to ask for waivers of environmental laws to protect mammals? Why or why not?

QUESTION 15-17. In November 2008 the U.S. Supreme Court ruled 5-4 that a lower court had acted improperly in denying the U.S. Navy's request to deploy and use high-intensity sonar to train its personnel. The Court's majority said that the Navy's need to train personnel trumped any adverse impact the activity would have on marine mammals, which the court majority asserted had not been conclusively demonstrated in any case. Research the issue and explain why the Court ruled as it did. Do you agree with the Court's 5-4 majority? Why or why not?

Tropical Marine Ecosystems

Coral Rocks! The Value of the World's Coral Reefs

KEY QUESTIONS

What are coral reefs? Where are they located, and what organisms contribute to reef construction and function?

How important are coral reefs as reservoirs of biodiversity?

How important are coral reefs to marine fisheries?

What is their significance in the carbon cycle?

What are the global threats to the health of coral reefs?

Coral reefs are increasingly under threat from human actions. Nearly half a billion people live within 100 km (62 miles) of coral reefs. *Status of Coral Reefs of the World: 2004*, by the Australian Global Coral Reef Monitoring Network,[1] documents the impacts humans have on the global coral reef crisis. In addition to natural forces with which corals have evolved for millions of years, the report lists these major anthropogenic (human caused) stresses to coral reefs:

- Sediment and nutrient pollution from the land
- Overexploitation and damaging fishing practices
- Engineering modification of shorelines
- Global climate change causing coral bleaching, rising sea levels, and potential acidification of shallow marine water threatening the ability of corals to form skeletons

This issue focuses on the importance of coral reefs, their role in global carbon balance, the effects of climate changes and other human impacts, and what can be done to protect them.

---

1. Wilkinson, C. (ed.). 2004. *Status of Coral Reefs of the World: 2004*. Australian Institute of Marine Science, Townsville, Queensland, Australia. 301 p.

• •

Background

Coral refers to marine invertebrate organisms belonging to the Phylum Cnidaria, Class Anthozoa, or to the hard, calcareous structures made by these organisms. An individual coral animal, known as a polyp (Figure I 16-1a and I 16-1b), resembles a minute anemone. It is the ability of these polyps to remove dissolved calcium carbonate from seawater and to deposit it as part of their rocky skeleton that forms coral reefs. How coral animals do this is not precisely known. What is known is that most corals need algae living in their tissues in order to precipitate the $CaCO_3$ from seawater. More about this, as well as information on coral reefs that do not require these algae, is discussed later here. These algae, known as zooxanthellae (pronounced zōh-zan-THEL-lee), are members of a group of single-celled organisms called dinoflagellates, which belong to the kingdom Protista (discussed later here and on page 71). Because both organisms benefit from the relationship (discussed later here), the relationship is known as mutualism.

Coral reefs represent some of the most important real estate on the planet. They cover approximately 600,000 km² (230,000 miles²) in the tropics, an area roughly comparable in size to the state of Texas. They are a major oceanic storehouse of carbon and may contain up to a million species of organisms, only a tenth of which have been identified.

FIGURE I 16-1A Coral polyps with their tentacles extended. Each individual polyp in a single coral head is genetically identical, making the colony especially susceptible to disease and environmental change.

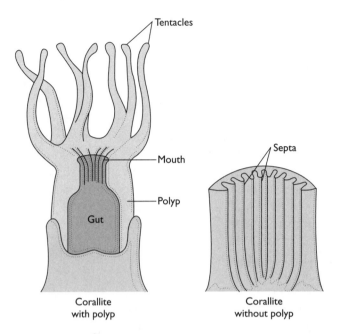

(B) CORAL STRUCTURE

★ **FIGURE I 16-1B** Diagram of a coral polyp within its calcareous skeleton, called corallite. The tentacles contain stinging cells called nematocytes.

Reefs have been built by organisms on and off for the past several billion years. Corals have been important contributors to reefs only for about the last 400 million years, and only within the past 50 million years have modern corals assumed their reef-building roles. Noncoral reefs built by cyanobacteria were present more than 3 billion years ago and thus are among the most ancient structures built by organisms.

Along with tropical rainforests, coral reefs, particularly those in the tropical Indo-Western Pacific Ocean, have the highest known biodiversities of any ecosystem on Earth. Globally, however, coral reefs are not prospering; indeed, their survival is threatened by nutrient pollution, sedimentation, overfishing, global warming, and even ecotourism. The severity of this problem from an economic viewpoint alone can be appreciated if you realize that tourists spend over $100 billion each year to visit locations near reefs. Florida reefs, for example, bring in at least $2 billion each year to that state's economy.

Coral Reefs and Fishing

Although reefs cover less than 0.2% of the ocean surface, they harbor a quarter of all marine fish species. Fish in coral communities are of two basic types: herbivores that feed on plants (algae and sea grasses) and carnivores that eat other animals.

Fishers have been plying reefs for millennia, and today, reefs provide food and employment for millions. During the past few decades, however, overfishing has degraded reef communities. Damage to Philippine reefs has resulted in the loss of 125,000 jobs. You may be able to anticipate some of the reasons for the degradation of coral reefs: If too many herbivores are removed, the marine algae that they eat may grow out of control and smother the coral. Removing carnivorous fish can also upset the reef's ecological balance. Can you see why?

Although removal of the fish may harm reefs ecologically, some methods of fishing, such as dynamiting, kill the hard coral colonies directly. Another method, called *muro-ami*, employs divers (typically young boys) who bounce rocks tethered to lines off the coral to herd fish. This method, which originated in Japan and is used in the Philippines, typically destroys about 17 m^2 of coral cover per hectare per operation. Typically, 30 muro-ami boats repeat an operation about 10 times a day. It may take 40 or more years for reefs that are destroyed by fishing practices to recover, if allowed to.

In many areas, muro-ami has been replaced with a technique called *pa-aling*, in which divers use compressed air instead of rocks to drive fishes out of coral heads and reefs.

Both muro-ami and pa-aling have been banned in the Philippines; however, the practices apparently continue there and in other areas.

Macroscopic marine algae grow amid and on the coral. Some of these algae are themselves carbonate producers and may be second only to corals in carbonate production. These algae, as well as noncalcareous algal species, may be eaten by parrotfish and other herbivores that find safety in the numerous nooks and crannies of the reef itself. These fish in turn become food for predators.

Fish, as well as invertebrates, are also removed from coral reefs. Fishers frequently use cyanide, which ideally stuns the animals but in fact kills many of them.

Conditions That Affect Reef Growth

Corals have evolved in a relatively stable environment and thus are adapted to a narrow range of environmental conditions, including light, temperature, salinity, sedimentation, and nutrient levels; therefore, they are vulnerable to even slight changes in these variables. The following discussion refers to shallow-water corals. Deep-water corals are discussed later.

■ Environmental conditions

- **Light**. Reef-building (hermatypic) corals are typically restricted to shallow waters ($<$ 50 m deep, but as deep as 75 m in the case of Pulley Ridge off Florida's south-west coast) because they require enough light for photosynthesis. Why would simple, eyeless, invertebrate animals such as corals need light? This requirement is due to the presence of zooxanthellae (discussed previously), which live within coral cells. These dinoflagellates, thousands of which live within the cells of a single polyp, require light to carry on photosynthesis. The high-energy end-products of photosynthesis (such as glucose) are used to nourish not only the zooxanthellae but are absorbed by the coral for their use as well. Some of the organic products are transferred to the coral polyp as a nutritional supplement to the food obtained when the coral feeds using its tentacles to entrap prey. Zooxanthellae also contribute oxygen and remove some waste materials, and they are involved in calcification. (Thus, the association between coral polyps and zooxanthellae is mutualistic, as introduced earlier.)

- **Temperature**. Shallow-water corals require a relatively constant, moderately high temperature. The global distribution of coral, in fact, correlates best with surface temperature: Corals are not generally found where winter surface water temperatures fall below 20°C, and they generally expel their algae and may die if water temperature exceeds 30°C. Thus, reefs are generally confined to waters between the Tropics of Cancer and Capricorn. This location protects them from low water temperatures but means that most corals live near their upper thermal limit of tolerance.

- **Salinity**. Average salinity cannot vary much from 35 parts per thousand (average seawater salinity) for corals to survive (except for specialized Red Sea corals, which are adapted to higher salinities).

- **Sedimentation**. Rates of sedimentation must be low and grain size relatively coarse or corals can be easily smothered. High levels of sediment in the water cause two problems: First, the corals' filtering apparatus can be clogged, making it difficult for them to feed on the tiny creatures that comprise their diet, and second, the sediment can blanket the colony and keep the zooxanthellae from photosynthesizing. Sedimentation, caused by runoff from construction sites, mining activities, industry, deforestation (e.g., in Haiti), and agriculture, is one of the reefs' most lethal enemies.

- **Nutrient levels** (i.e., the concentrations of phosphorus and nitrogen compounds). Nutrient levels must also be low for coral to survive. In fact, most coral reefs flourish in nutrient "deserts", or, more precisely, "oligotrohic" systems (from the Greek *oligo*, meaning little or few, and *trophe,* meaning nutrients, food). High nutrient concentrations can stimulate the growth of algae, which can smother the corals. Expansion of intensive, Western-style agriculture on land, with its massive doses of water-soluble, nutrient-rich fertilizer may lead to die-offs in offshore reefs, and in the United States, treated, nutrient-rich sewage from Florida's West Coast may be among sources of degradation affecting the once-healthy coral reefs in the Florida Keys.

Reefs and the Carbon Cycle

Coral reefs are an integral part of the planet's carbon cycle and may help control the earth's surface temperature. The living polyps of a coral reef represent only a thin film covering a massive rock skeleton built up over centuries or millennia by generations of coral polyps. This hard foundation is made of calcium carbonate, which is derived from calcium ions and carbon dioxide dissolved in seawater. As the coral polyps grow, they (along with calcareous algae) precipitate calcium carbonate, which supports the growing colony of coral and helps to cement coral rubble together.

Corals are extremely efficient at fixing carbon (remember that we are talking about the carbon cycle and that carbon is a key component of calcium carbonate). Each kilogram of $CaCO_3$ contains almost 450 grams (1 pound) of carbon dioxide.

[Geologists, by the way, believe that the Earth's early atmosphere was greatly enriched in carbon dioxide, mainly from volcanic gases. As life evolved, some marine bacteria began to precipitate (solidify) calcium carbonate, in effect storing or fixing carbon dioxide, removing it from the atmosphere. Today, immense masses of carbonate sedimentary rock on continents attest to the enormous amount of carbon dioxide removed from the atmosphere. If all of this CO_2 were returned to the atmosphere, the Earth's surface temperature might approach that of Venus, where surface temperatures (460°C) are hot enough to melt lead.]

Coral Bleaching

One of the severest threats to coral survival in recorded history is called "bleaching," and it happens when corals expel the zooxanthellae living in their tissue. The greatest stress seems to come from high water temperature, one consequence of global warming. Data from the past century suggest that such bleaching events could become more severe and persistent as the planet warms. Warming compounds the stresses on reefs from local human-induced sources that we described earlier, but could extend the corals' range.

What is the future of coral reefs? Unless humans take these threats seriously and act together to mitigate them, the future is not bright.

How Fast Are Coral Reefs Disappearing?

In this analysis, you will perform calculations that allow you to project when coral reefs would virtually disappear from the planet.

Use this information in the following questions:

Radius of Earth:	6,378 km
Total area of world's oceans:	360×10^6 km^2
Total area covered by shallow-water coral reefs (approx.):	$\geq$600,000 km^2
Total area of ice-free land:	1.31×10^{14} m^2

First, let us see what percentage of the planet coral reefs cover.

QUESTION 16-1. Using the information provided above, calculate the area of earth's surface.

(Recall that we are assuming that the earth is a perfect sphere. The formula for calculating the area of a sphere is this: area = $4\pi \times (r)^2$, where π = 3.1416, r = radius).

QUESTION 16-2. If coral area is approximately 600,000 km^2, what percentage of the earth's surface is covered with coral reefs?

More than 27% of the world's reefs may have already been lost. Indonesia, Australia, and the Philippines possess the largest area of remaining coral reefs. Researchers John Bruno and Elizabeth Selig of the University of North Carolina at Chapel Hill examined coral reef loss from 1968 to 2004 and concluded that coral reef cover declined 1% annually, which is higher than the 0.4% rate of disappearance of tropical rainforest.[2]

QUESTION 16-3. Use the doubling time formula (70/r, see Appendix) to calculate when coral reef coverage will be 300,000 km^2, assuming a 1% rate of decrease.

For any entity, such as the rate of coral reef loss, to decrease by a factor of 1,000 (actually 1,024), it takes 10 doubling periods.

QUESTION 16-4. Using the doubling time you just calculated, determine when coral reef coverage would be reduced to 1/1,000 of its current coverage, or an impossible 600 acres.

If the rate of reef loss increases with increasing human population, the impact will be much more severe. Your calculations should have shown, even if we double the present rate of reef loss, that there is still adequate time to save this irreplaceable ecosystem, given the will to act.

These numbers may be misleading because even though the total coverage of reefs is decreasing at a small rate, the amount of coral at each reef may be decreasing more rapidly. Bruno and Selig found that only 2% of Indo-Pacific reefs have the same amount of live coral as they did in the 1980s, and the remaining corals were likely less healthy and diverse.

A key assumption in your previous calculations was that the rate of destruction remains constant; however, global climate change will likely increase surface seawater temperature. Corals, like other organisms, live closer to the maximum temperature they can tolerate (known as the critical thermal maximum) than to the minimum. Thus, it is possible that once this threshold is reached, the rate of bleaching and death of corals may accelerate. Additionally, acidification of ocean waters, a likely consequence of growing atmospheric carbon dioxide, can reduce the ability of coral animals to form their calcium carbonate reefs.

QUESTION 16-5. Extinction is forever. Would you be willing to let some species of corals disappear forever? All species? Discuss.

QUESTION 16-6. Who is responsible, if anyone, for ensuring the survival of coral reefs? Explain your answer.

QUESTION 16-7. Identify actions that could be taken to help coral reefs recover and protect them. Which are most practical? What are the impediments to implementing these actions?

---

2. Bruno, J. F., and E. R. Selig. 2007. Regional Decline of Coral Cover in the Indo-Pacific: Timing, Extent, and Subregional Comparisons. PLoS ONE 2(8): e711 doi:10.1371/journal.pone.0000711.
   Available at http://www.plosone.org/article/fetchArticle.action?articleURI=info:doi/10.1371/journal.pone.0000711.

QUESTION 16-8. Discuss whether you think promoting ecotourism, nature-based tourism that appeals to environmentally oriented individuals, would contribute to solving the coral reef problem, or instead might worsen the problem.

Deep-Water Corals

Not all coral reefs are shallow. Unknown until about 200 years ago, deepwater coral reefs, also called coral banks or cold water corals, live as deep as 1,000 m. These coral reefs, in many cases as diverse as their shallow water counterparts, consist of live coral on top of coral rubble or unconsolidated sediment.

Deepwater corals exist worldwide seaward of the continental shelf, including along the U.S. east coast. Stetson reef, about 140 miles east of Charleston, SC, covers over 2,000 mi^2 of sea floor. Deep-water coral reefs are only now being studied, in part because of the expense of reaching them. Unfortunately, trawling and dredging (see Issue 20) threaten to destroy these ecosystems, which are particularly vulnerable to disturbance because the corals are very slow growing and fragile.

Actions are being taken to save these unique systems from damage from fishing. The South Atlantic Fishery Management Council has proposed a plan to restrict bottom trawling from four areas where deep-sea corals are present. Unfortunately, there are renewed calls for increased offshore drilling, yet another major threat to these systems.

Media Analysis

Go to www.npr.org, and search for "Caribbean Coral Reefs Dying at Dramatic Rate."[3] Listen to the 4-minute report from 2003, and answer the following questions.

QUESTION 16-9. What is different about this *Science* magazine report compared with other studies documenting coral reef decline?

QUESTION 16-10. What did U.S. Geological Survey marine ecologist Caroline Rogers say about changes in coral reefs in the U.S. Virgin Islands over the last 25 years?

QUESTION 16-11. Why do you think Rogers considers it to be very unusual for a marine invertebrate like coral to be considered for the endangered species list?

QUESTION 16-12. List reasons presented in this report why coral reefs are declining.

QUESTION 16-13. Explain how overfishing harms coral reefs.

QUESTION 16-14. What did researcher Isabelle Cote discover about the decreases in coral cover?

---

3. Or go directly to http://www.npr.org/templates/story/story.php?storyId=1340551.

QUESTION 16-15. Do you think the description that ecosystems are "less colorful and vibrant" conveys the seriousness of the threat to coral reefs? Explain your answer.

QUESTION 16-16. With what does researcher Cote compare the rate of destruction of coral reefs?

QUESTION 16-17. Explain whether you think the plight of tropical forests or other ecosystems is more in the public eye than that of coral reefs.

QUESTION 16-18. Explain whether you think that this study will be a "wake-up call" to authorities to take these man-made problems seriously.

QUESTION 16-19. Explain why you think authorities in power have not taken these problems seriously in the past.

QUESTION 16-20. Identify some of the problems the U.S. Virgin Islands might experience if coral reefs disappeared from their waters.

QUESTION 16-21. How might strict application of the Precautionary Principle (see pp. 5–6) help protect coral reefs?

Media Analysis 2

Go to www.npr.org, and search for the 2008 report "World's Coral Reefs Face Renewed Threats."[4] Then listen to the 3-minute report, and answer the following questions.

QUESTION 16-22. What event turned over 90% of Jamaica's coral reefs into "underwater ghost towns?"

QUESTION 16-23. What is the purpose of the "Year of the Reef?"

QUESTION 16-24. What is meant by "paper reef-protection zones?"

QUESTION 16-25. How is the United States described by the Australian marine biologist Clive Wilkinson? Explain what he meant.

---

4. Or go directly to http://www.npr.org/templates/story/story.php?storyId=18456171.

W(h)ither the Mangroves?

KEY QUESTIONS

What are mangroves?

Where are mangroves found?

What is the global state of mangroves?

What are threats to the health of mangroves?

Mangroves are tropical salt-tolerant trees that grow at the water's edge, between 32° N and 38° S latitude. Five countries—Indonesia, Australia, Brazil, Nigeria and Mexico—are home to almost half of the entire global area of mangrove ecosystems.

Florida, has three species: red mangrove (*Rhizophora mangle*), black mangrove (*Avicennia germinans*), and white mangrove (*Luguncularia racemosa*). The mangrove ecosystem is known variously as a mangal, mangrove forest, or mangrove swamp.

In some species, the leaves of mangroves can rid the plant of salt it absorbs at its roots; in others, the roots prevent salt from entering. As a result of these two methods of handling salt, too much of which can damage the trees' tissues, mangroves can grow in ocean water, brackish estuaries, and salty soil.

Mangroves are unique in that they produce "stilt roots," which project above water, thus allowing the tree to absorb oxygen and deliver it to the submerged or buried part of the root system. This is important because the trees typically grow in muddy hypoxic (low-oxygen) conditions (Figure I 17-1).

Stilt roots of mangroves at the water's edge form complex networks, trapping sediment and preventing its transport further seaward. They also provide a myriad of hiding places for a great diversity of marine life, especially the juvenile and larval forms of fish, crustaceans, and mollusks. Mangroves flourish as coastal marine tidal forests and as of 2007 covered about 15 million hectares, down from nearly 18 million in 1980. According to the Food and Agricultural Organization of the United Nations, approximately 100,000

FIGURE I 17-1 Mangrove trees, showing stilt roots above and below water. Describe briefly why mangroves are so effective as sediment traps.

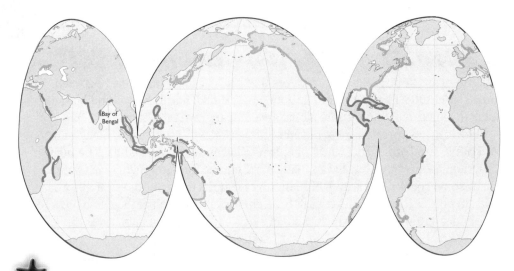

FIGURE I 17-2 Global distribution of mangrove swamps. Note the small area of mangroves at the northern tip of the Bay of Bengal. This area in India and Bangladesh constitutes the largest contiguous area of mangroves in the world. Areas of mangrove along the NE coast of Brazil and the coast of Ecuador are being destroyed to make way for industrial shrimp farming.

hectares (247,000 acres) of mangroves were lost annually between 2000 and 2005, fewer than in previous decades.

Mangrove communities flourish mainly along the shores of estuaries, along coasts, and at the mouths of large rivers like the Irrawaddy in Myanmar. Figure I 17-2 shows the global distribution of mangrove swamps.

These locations mean that mangrove communities tend to be at risk from development, especially in countries with few if any land-use controls. Mangrove environments are readily cleared for aquaculture, for resort developments, and for housing. They are also threatened globally by rapid sea-level rise and locally by agricultural runoff, oil exploration and development, deforestation for biomass fuel, and gas exploration and production, the latter for example, along the Persian Gulf and the coasts of Australia and Nigeria.

The southwest coast of Florida contains the most extensive mangrove forests in the continental United States, and yet even there, mangrove forests continue to decline because of development.

Aquaculture development along tropical coasts directly threatens the survival of mangrove forests (discussed later here), with the possible exception of protected areas such as Australia's Great Barrier Reef, the world's largest reef. Industrial shrimp farming along the northeast coast of Brazil, for example, puts at risk much of that nation's one million surviving hectares of mangroves.

Mangroves and Coral Reefs

Mangrove and coral reef communities are sometimes cited as examples of "linked habitats" because of their close association. Mangroves are often found landwards of coral reefs. Although mangroves grow in quiet, sediment-rich waters, coral reefs require clear waters free of fine sediment that can smother the corals and kill the reef. The communities are also complimentary in that coral reefs protect mangroves from wave erosion and storm damage whereas mangroves provide shelter for juvenile stages of numerous reef animals. The removal of mangroves often leads to the degradation of even disappearance of the reef.

■ Focus: Shrimp Mariculture in Ecuador

Mangrove habitat is often destroyed along tropical coastlines to support construction of artificial ponds for shrimp aquaculture. This activity has been responsible for the loss of two thirds of mangrove forests in the Philippines, more than half of Thailand's mangroves, and more than a fourth of Ecuador's (one fifth of Ecuador's mangrove forests were cleared for shrimp mariculture between 1979 and 1981 alone!) (Table I 17-1).

TABLE I 17-1	Area (in hectares) of mangroves and shrimp ponds in Ecuador, 1969–1999					
Year	1969	1984	1987	1991	1995	1999
Mangrove area (thousand ha)	203.695	182.157	175.157	162.186	146.938	149.556
Shrimp pond area (thousand ha)	0	89.368	117.728	145.998	178.071	175.253

*http://www.clirsen.com/

Ecuador's shrimp exports—60% of which come from artificial ponds created by removing mangroves—were second only to oil in the early 1990s as a producer of foreign exchange. Indeed, Ecuador is the largest supplier of imported shrimp to the United States.

Local artesanos (artisanal fishers) collect juvenile shrimp from the mangrove-fringed estuaries and beaches and transfer them to ponds from which mangroves have been cleared. These postlarval shrimp spend 3 to 5 months in the mangrove-fringed estuaries, which provide a nutrient-rich environment ideal for fast growth and also offer protection from natural predators. The extent of the damage resulting from removal of the juveniles to both the shrimp population and the ecosystem is not yet known.

As mangroves were destroyed, natural populations of shrimp declined as well.

In 1975 and 1985, laws were passed outlawing the conversion of mangrove forests to shrimp ponds; however, the laws were not enforced and mangrove deforestation continued at a rate of approximately 3,000 hectares annually. Demand for shrimp remains high.[1]

QUESTION 17-1. How is the preservation of the mangrove environment related to the viability of shrimp mariculture in Ecuador?

QUESTION 17-2. How does the mangrove environment protect the postlarval stage of shrimp?

QUESTION 17-3. Table I 17-1 shows the area occupied by mangroves and shrimp ponds from 1969 to 1999. On the axes below, or on our website (http://www .jbpub.com/abel/), graph the mangrove area for the period 1969 to 1999. On the same axes, graph the growth of shrimp ponds. Interpret the trends. What additional evidence would you like to have to establish a cause–effect relationship between loss of mangroves and growth of shrimp ponds?

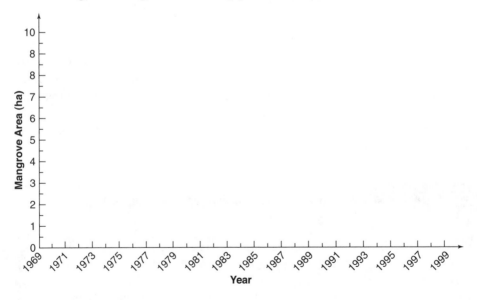

---

1. The full story is available at www.colby.edu/personal/t/thtieten/aqua-ecua.html.

Mangroves and Tsunamis

On December 26, 2004, a massive earthquake struck the south coast of the island of Sumatra, Indonesia, killing 250,000+ people in Indonesia, Thailand, India, Sri Lanka, and elsewhere around the Indian Ocean. It was one of the greatest natural disasters in human history. Massive water waves known as tsunami, generated when the seismic energy released by the earthquake entered the ocean, were responsible for the deaths and destruction.

At the location of the tsunami's maximum intensity, the force of the waves was so powerful that structures and people stood little chance.[2] Along Sumatra's coastline, tsunami waves may have reached heights of 15 to 30 m over a distance of 100 km. Further away, however, according to a report in *Science* magazine,[3] "areas with coastal tree vegetation were markedly less damaged than areas without." The trees referred to were mangroves, and they absorbed much of the energy of the waves, protecting the area landward of the mangrove forest. According to the report, mangrove area had been reduced before the tsunami (between 1980 and 2000) by 26% in the five countries most affected by the tsunami, from 5.7 to 4.2 million hectares (14.1 to 10.4 million acres) because of human activities that included aquaculture and development of resorts.

Neil Burgess, a co-author of the study, stated[4]: "Just as the degradation of wetlands in Louisiana almost certainly increased Hurricane Katrina's destructive powers, the degradation of mangroves in India magnified the tsunami's destruction. Mangroves provide a valuable ecological service to the communities they protect."

Mangroves and Sharks

A new study[5] concludes that the development of a resort in Bimini, Bahamas that involved dredging and removal of mangroves significantly damaged the nursery grounds of lemon sharks (*Negaprion brevirostris*; Figure I 17-3 and cover). The development, which started in 1997, was projected to include a 930-room hotel, a casino, 18-hole golf course and two large marinas. By the end of 2006, approximately 750,000 m³ had been dredged from the main lagoon, and approximately 0.8 hectares (2 acres) of mangroves forest had been cleared.

QUESTION 17-4. The mangrove forest that was cleared represented 30% of the total mangrove area fringing the North Sound of Bimini. Calculate the mangrove area before clearing, as well as the area of mangroves remaining by the end of 2006. Report your answer in hectares (ha) and acres (ac) (1 ac = 0.4047 ha).

---

2. For a vivid video of the force of the tsunami, go to http://www.usc.edu/dept/tsunamis/2005/tsunamis/041226_indianOcean/sumatra/tsunami_video_popup.html.
3. Danielsen, F., et al. 2005. The Asian Tsunami: A Protective Role for Coastal Vegetation. *Science* 310:643.
4. As quoted in Mangroves Shielded Communities Against Tsunami. *ScienceDaily* (October 28, 2005). Available at http://www.sciencedaily.com/releases/2005/10/051028141252.htm.
5. Jennings, D. E., et al. Effects of large-scale anthropogenic development on juvenile lemon shark (*Negaprion brevirostris*) populations of Bimini, Bahamas. *Environmental Biology of Fishes* (in press).

FIGURE I 17-3 Young lemon shark (*Negaprion brevirostris*) in a mangrove nursery. [Photo by D. C. Abel.]

Although the total area of mangroves around the North Sound may seem small, this habitat was one of the most important lemon shark nurseries in the northwest Bahamas.

Understanding the biology of the lemon shark has been a major focus of shark biologist Dr. Samuel Gruber of the Bimini Biological Field Station. Research conducted by Gruber and his co-workers over approximately 2 decades has shown that mangroves and the lagoon fringed by the mangroves are critical to the growth and survival of juvenile lemon sharks during their first several years of life, after which they move offshore. Young lemon sharks take advantage of the protection from predation offered by the mangroves (one of their most important predators is larger lemon sharks), as well as the abundant food supply in the lagoon. The continued viability of this area as a nursery ground is essential to the survival of lemon shark populations in the area.

Their study of lemon sharks showed that after the development started survival rates in lemon sharks decreased, growth rates decreased, and sharks remaining in the area were less healthy than comparable sharks in undisturbed areas.

QUESTION 17-5. Identify ways in which the loss of mangroves may have caused the changes in lemon sharks.

The study further showed that dredging and removal of the mangroves caused changes in the structure of the North Sound habitat, including a decline of nearly 18% in the cover of seagrass (*Thallasia testudinum*).

It is difficult to determine whether the dredging or the removal of the mangroves caused the most damage because the survival of the mangroves was also negatively affected by the increased tidal flow rates, sedimentation, and salinity associated with the dredging.

QUESTION 17-6. The development's websites[6] state that they have instituted "a habitat creation and restoration program that will maintain the surrounding mangrove wetlands. Together with marine biologists and governmental organizations, we … are committed to the preservation of the natural wonders of Bimini for generations to come." Analyze the statement in light of the scientific results presented previously here.

QUESTION 17-7. Discuss whether you think that protecting the nursery grounds of lemon sharks would be sufficient grounds for terminating the project, which is well-underway.

Media Analysis 1

Go to www.npr.org, and search for "Gas Flaring Disrupts Life in Oil-Producing Niger Delta"[7] from 2007. Listen to the nearly 9-minute program, and read and study the transcript and pictures. Then answer the following questions. (Note that the discussion of mangroves is late in the report.)

QUESTION 17-8. Natural gas is a premium fuel in great demand all over the world for home heating and electric power generation. It produces only half the greenhouse gases of coal when burned, and when processed, none of the toxic sulfur gases. What do you think the author means when he describes the burned-off (flared) natural gas as "unwanted?"

QUESTION 17-9. Describe the geography of the region in Nigeria where the oil is extracted and pumped to shore.

QUESTION 17-10. As of 2007, how much gas was being flared in Nigeria annually, and how does this amount compare to energy demands of the African continent?

QUESTION 17-11. Describe the physical environment those who live close to the gas flares experience daily.

QUESTION 17-12. How much carbon dioxide was being emitted each year, as of 2007, by gas flaring?

QUESTION 17-13. How much has Nigeria reduced its gas flaring since the 1980s?

QUESTION 17-14. Speculate on the impact oil development is having on the mangrove communities of the Niger Delta.

---

6. www.biminibayresort.com/about-preservation.htm and www.biminibayfacts.com/environment.html.
7. Or go directly to www.npr.org/templates/story/story.php?storyId=12175714. Also, see *National Geographic* Curse of the Black Gold: Hope and Betrayal in the Niger Delta by Tom O'Neill. Available at http://ngm.nationalgeographic.com/2007/02/nigerian-oil/oneill-text.

. . . .

Media Analysis 2

Go to www.npr.org, and search for "'Times' Series Links Ocean Health, Global Warming."[8] Listen to the 5:11 minute program from mid 2006, and then answer the following questions.

QUESTION 17-15. How did California Governor Schwarzenegger characterize the "controversy" over global warming/climate change?

QUESTION 17-16. What is the consensus over toxic algae in the oceans referred to by the *Times* reporter?

QUESTION 17-17. What frustrations did scientists studying climate change express to the reporter?

QUESTION 17-18. Do you agree with the frustrations expressed by the climate scientists? Why or why not? Is your conclusion grounded in critical thinking? Explain.

. . . .

Media Analysis 3

Go to www.npr.org, and search for "Tracking Climate Change in Senegal."[9] Listen to the first 8 minutes of the 2008 program, and then answer the following questions.

QUESTION 17-19. Instead of HIV/AIDS or malaria, what was the "real story" journalist Bob Butler found in Senegal?

QUESTION 17-20. Describe the impacts of the "torrential rain" event Butler experienced.

QUESTION 17-21. What explanations did experts offer to Butler for the changes he observed in Senegal?

QUESTION 17-22. Describe the effects overfishing is having on Senegal, especially the impact of factory ships from Europe.

QUESTION 17-23. What services do the mangroves provide to the Senegal fisheries?

QUESTION 17-24. Why are the "most knowledgeable" climate change experts worried, according to Butler?

---

8. Or go directly to www.npr.org/templates/story/story.php?storyId=5597652.
9. Or go directly to www.npr.org/templates/story/story.php?storyId=18916450.

Media Analysis 4

Go to www.npr.org, and search for "Shrimp Farming."[10] Listen to the 13-minute program from 2001, and then answer these questions. (You will need Real Player to listen to the program.)

QUESTION 17-25. What percentage of the shrimp consumed in the United States in 2001 was from shrimp "farms"?

QUESTION 17-26. What nation is the world's biggest consumer of shrimp?

QUESTION 17-27. What ecosystem was removed to make room for Ecuador's shrimp "farms?"

QUESTION 17-28. Explain how the shrimp "farms" impact the mangroves, and describe the permanency of the ponds.

QUESTION 17-29. How has the shrimp "farming" industry affected the catch of the local shellfish gatherers?

QUESTION 17-30. Is the reported destruction of shrimp farms by local activists and shellfish farmers a legitimate form of protest? Why or why not?

QUESTION 17-31. What industry was described as the "perfect example of globalization run amok?"

QUESTION 17-32. Shrimp farming was described as second only to what crop in profitability?

QUESTION 17-33. Describe the attitude of local judges to cases involving mangrove destruction.

QUESTION 17-34. What percentage of Ecuador's population was described by the shrimp farm businessman as being dependent on the shrimp industry?

QUESTION 17-35. What was the impact of the recently imported virus on shrimp farms, and from where did the virus come?

QUESTION 17-36. In summary, what effect is the shrimp farming industry having on the local fishing industry?

---

10. Or go directly to www.npr.org/templates/story/story.php?storyId=1118568.

PART 8

Fisheries and Aquaculture

Catch of the Day: The State of Global Fisheries

KEY QUESTIONS

What is the state of global fisheries?

What are the trends in "harvesting" of wild marine fish?

What is bycatch?

What is the environmental impact of commercial fishing?

How much seafood do we eat?

How important is seafood as a protein source?

Is "sustainable fisheries" an oxymoron?

By virtually all accounts, global fisheries are at a crisis. Few fisheries are well managed, and overfishing, coastal habitat destruction, pollution, and climate change threaten the remainder.

A 2004 report by the U.S. Commission on Ocean Policy reported that "25% to 30% of the world's major fish stocks are overexploited and that US fisheries are experiencing similar difficulties." The report further found that about a quarter of the nation's 267 major fish stocks are either already overfished or are experiencing overfishing. A 2004 report by the Royal Commission on Environmental Pollution concludes that fishing should be banned in nearly a third of U.K. waters. In the northeast Atlantic and adjacent seas, more than 40% of commercial fish species are being fished at higher than sustainable rates. "Because it happens at sea, the damage is largely hidden," Chairman Sir Tom Blundell said.

The day following publication of the Royal Commission report, the European Union proposed a major series of fishing cuts, including closure of heavily depleted cod grounds in the North Sea, the Irish Sea, and off the west coast of Scotland. They also mandated reductions of up to 60% in herring catches, 34% for cod, and 27% for mackerel.

Europe's fisheries commissioner, Joe Borg, said the proposals were "a balance between what is biologically necessary and what is economically reasonable."

The Food and Agricultural Organization of the United Nations (FAO) reported[1] that, as of 2005, only

Around one-quarter of the stock groups monitored by FAO were underexploited or moderately exploited and could perhaps produce more, whereas about half of the stocks were fully exploited and therefore producing catches that were at, or close to, their maximum sustainable limits, with no room for further expansion. The remaining stocks were either overexploited, depleted, or recovering from depletion and thus were yielding less than their maximum potential owing to excess fishing pressure.

Over 2.6 billion people depend on fish for protein. The top capture fishery countries are China, Peru, Japan, the United States, Indonesia, Russian Federation, India, Thailand, Norway, Iceland, and the Philippines.

QUESTION 18-1. Do you think that the U.S. government, or the fishing industry, "manages" fisheries? If you answered yes, discuss whether it is your impression that the United States manages fisheries in a way that strikes "a balance between what is biologically necessary and what is economically reasonable." If not, do you think fisheries should be managed? Why?

Have We Reached "Peak Fish"?

Fish and shellfish are obtained in three ways: marine capture fisheries, fresh water capture fisheries, and aquaculture, or fish and shellfish "farms" (see Issue 21). According to the FAO, the top five capture fish species (by weight) are

- Peruvian anchoveta (*Engraulis ringens*), 10.7 million tonnes in 2004, used primarily for fish meal and oil

- Alaskan pollock (*Theragra chalcogramma*), 2.7 million tonnes, used for fish fillets, including fast food fish sandwiches

- Blue whiting (*Micromesistius poutassou*), 2.4 million tonnes, used in production of fish sticks and other portioned fish products

- Skipjack tuna (*Katsuwonus pelamis*), 2.1 million tonnes, primarily eaten fresh, frozen, or canned

- Atlantic Herring (*Clupea harengus*), 2.0 million tonnes, eaten cooked, raw and canned

In the United States, the two largest catches are pollock and menhaden.

Table I 18-1 shows catch data for Peruvian anchoveta, Atlantic pollock, and skipjack tuna.

---

1. Available at http://www.fao.org/docrep/009/A0699e/A0699e00.htm.

TABLE I 18-1	Catch data (million tonnes) for selected species, 1990–2006								
Species	1990	1995	2000	2001	2002	2003	2004	2005	2006
Peruvian Anchoveta	3.77	8.64	11.28	7.21	9.70	6.20	10.68	10.22	7.01
Atlantic Pollack	5.74	4.81	2.94	3.14	2.65	2.89	2.69	2.79	2.86
Skipjack Tuna	1.28	1.66	1.97	1.83	2.04	2.18	2.10	2.36	2.48

*From FAO – Fisheries and Aquaculture Information and Statistics Service

QUESTION 18-2. On the axes below, or on our website (http://www.jbpub.com/abel/), graph the catch data for the three species in Table 18-1. What trends did you observe?

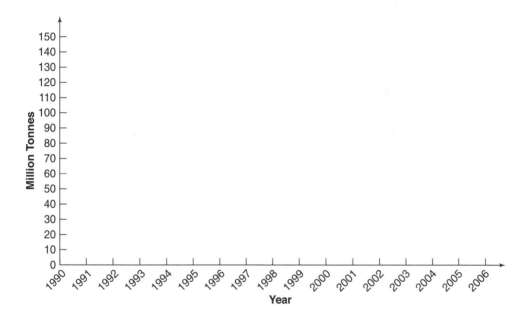

Catch Data for Peruvian anchoveta, Atlantic pollack, and skipjack tuna, 1990–2006.

In terms of economic value, shrimp are most important, followed by species known as groundfish (hake, cod, haddock, and Alaska pollock), tuna, and salmon.

In 2007, capture fisheries and aquaculture in combination produced about 143 million tonnes. Aquaculture comprised 36% of the total.

Table I 18-2 shows capture and aquaculture production data (in million tonnes) for marine waters from 1950 to 2007. For 2005, 81.6% of marine fishery production was wild caught. Only 24.9% of inland (i.e., freshwater) production was from capture fisheries in 2005.

TABLE I 18-2	World Fish Catch, Aquaculture, and Total Harvest 1950–2007			
Year	World Population (billions)	Total Wild Catch (million tonnes)	Aquaculture (million t)	Per Capita Wild Fish Production (kg. person^{-1})
1950	2.556	19.2	1.5	
1955	2.780	26.4	2.1	
1960	3.039	36.4	3.0	
1965	3.345	49.0	3.7	
1970	3.707	58.2	3.6	
1975	4.086	62.4	4.1	
1980	4.454	67.0	5.2	
1985	4.851	79.1	7.7	
1990	5.277	85.9	13.1	
1995	5.682	92.0	24.4	
2000	6.156	95.5	35.5	
2005	6.447	93.8	47.8	
2006	6.555	91	50	
2007 (projection)	6.625	91	52	

*Compiled by Worldwatch Institute (http://www.worldwatch.org) from U.N. Food and Agriculture Organization Yearbooks

QUESTION 18-3. On the axes below or on our website (http://www.jbpub.com/abel/), plot three separate lines: aquaculture production, capture production, and total production (which you must first calculate) for the years 1950–2007.

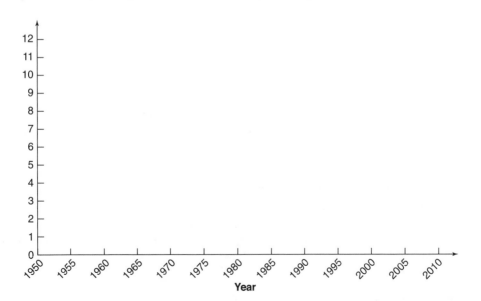

Aquaculture Production, Capture Production, and Total Production, 1950–2007.

QUESTION 18-4. Calculate the percentage increase in aquaculture, capture, and total marine production for the 16-year period from 1990 to 2007.

QUESTION 18-5. What do you conclude from your plots and your calculations?

The late Julian Simon in his 1996 book *The Ultimate Resource 2* (p. 104) stated, "No limit to the harvest of wild varieties of seafood is in sight…. It is still rising rapidly."

A July 13, 2003 press release from the U.S. House Committee on Natural Resources, stated this: "We know, contrary to the conclusions of the recent Pew Oceans Commission report, that there are indeed still plenty of fish in the sea."

QUESTION 18-6. Are the previous statements consistent with the trends you just observed and the conclusions of the U. S. Commission on Ocean Policy report? Explain your answer.

For the remainder of this issue, we focus on capture fisheries only. Issue 21 discusses the production and impacts of aquaculture.

In our opinion, the absolute tonnage of capture fishery production is less important than the amount per person (per capita). Do you agree?

QUESTION 18-7. Calculate the per capita fish supply (i.e., the number of kg per person) from 1950 to 2007, and then fill in the appropriate column in Table I 18-2. Recall that 1 tonne = 1,000 kg.

QUESTION 18-8. On the axes below, or on our website (http://www.jbpub.com/abel/), plot a graph of per capita fish supply (in kg/person) for the period 1950 to 2007.

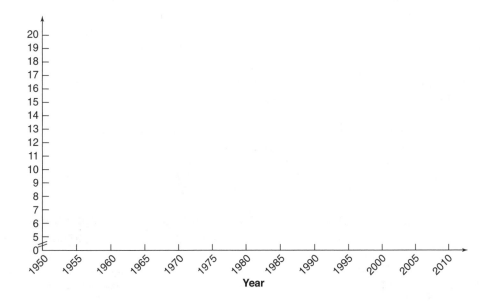

QUESTION 18-9. Describe any trends in your graph.

QUESTION 18-10. With global population growing at about 1.15% per year, do you think that capture fisheries can supply ever-increasing demand on fish stocks of a burgeoning human population? Have we reached "peak fish" production? Justify your answer. What assumption(s) did we use in formulating this question?

QUESTION 18-11. Suggest and discuss ways to make fisheries sustainable.

QUESTION 18-12. Do you think that the total fish production was equitably distributed globally. In other words, do you think that fishery production moved more in the direction of economic demand or nutritional need? Explain your answer.

Only a portion of fishery production is for human consumption. Fish and fish by-products, representing as much as 25% of wild-caught fish, are a mainstay of the pet food industry and are used as a constituent of animal feed (including aquaculture) as well.

Environmental Costs of Fishing

Much commercial fishing is environmentally destructive. The declines in fisheries catch that you calculated previously here mean real decreases in population sizes of numerous species that, in turn, may have repercussions for entire ecosystems. Additionally, commercial fishers are increasingly engaged in what has been called "fishing down the food web." This means that large predatory fishes are targeted first. These typically are the biggest fish in marine ecosystems, but they are also the least numerous. A study in the journal *Nature*[2] reported that 90% of large oceanic fish have been removed from the oceans. Although the magnitude of the decrease has been debated, fishery scientists are in agreement that there has been a drop and that it has been large.

The late Ransom Myers (Dalhousie University), lead author of the study, stated this:[3]

> *From giant blue marlin to mighty bluefin tuna, and from tropical groupers to Antarctic cod, industrial fishing has scoured the global ocean. There is no blue frontier left. Since 1950, with the onset of industrialized fisheries, we have rapidly reduced the resource base to less than 10%—not just in some areas, not just for some stocks, but for entire communities of these large fish species from the tropics to the poles.*

The ecological impact of the disappearance of large fishes is difficult to measure. One possible outcome is known as a trophic cascade, referring to changes in

---

2. Myers, R. A., and B. Worm. 2003. Rapid worldwide depletion of predatory fish communities. *Nature* 423:280–283.

3. Quoted in http://news.nationalgeographic.com/news/2003/05/0515_030515_fishdecline.html.

population levels of organisms at lower trophic levels resulting from changes in the population of the predators at or near the top of the food chain. A classic example is the overfishing of cod in the Gulf of Maine brought about by the advent of mechanized fishing methods in the 1930s. The decline of the predatory cod reduced predation pressure on their prey, smaller fish such as sculpins. As populations of sculpins and other small fish with similar ecological roles increased in number, populations of *their* prey decreased, ultimately resulting in huge population increases of sea urchins. With few controls on their population, these sea urchins consumed and thus eliminated entire kelp forests.

Jeremy Jackson, an internationally renowned marine ecologist at Scripps Institution of Oceanography at the University of California, San Diego, confirms that scientists do not yet understand the ramifications of overfishing: "What happens when we take away parts of the food chain is a big scientific research question. And nobody knows all the answers to that…. And it's pretty clear that we can't control the outcome."

Moreover, understanding the environmental and ecological impact of overfishing is made even more difficult because there may be a significant time lag between when overfishing occurs and when the ecological impact becomes evident.

When population levels of large fishes dropped to the point that they were no longer caught consistently or the expense of capture became prohibitive, commercial fishes switched to smaller fishes lower on the food web and to even lower trophic levels, including many more invertebrate species. Over time, this process has resulted in a decline in the average trophic level that is being fished. "Eating low on the food chain" is desirable (see Chapter 7), except when it results from overfishing of organisms higher on the food chain.

Also, as populations of species most easily captured (the "low-hanging fruit" of the oceans) have declined, commercial fishers have moved farther from shore and have increased the depth of capture.

QUESTION 18-13. Discuss whether you think commercial fishers' move to these new offshore and deeper environments is a good idea.

In addition to endangering the target species, many fishing methods destroy habitat (see Issue 20).

Finally, the environmental cost of commercial fishing extends to the pollution associated with the use of equipment (like fishing boats) and supplies, fuel spills, and transportation and refrigeration of fishery products. See issue 9 for the impact of "bunker fuel" used by marine vessels.

QUESTION 18-14. Capture fisheries have been described as "a race to catch the last fish." Discuss whether you agree or disagree with this assessment.

QUESTION 18-15. Sylvia Earle, a *National Geographic* Explorer-in-Residence and former director of the U.S. National Oceanic and Atmospheric Administration, has said, "People who want to make a difference can choose not to eat fish that are more important swimming alive in the ocean than swimming in lemon slices and butter." Do you agree? Why or why not?

QUESTION 18-16. Many nations provide heavy subsidies to their fishing industries. Discuss what you think the impact of rising energy (i.e., fuel and refrigeration) costs might be on world fisheries.

In the United States, Wild Oats Markets, a nationwide chain of more than 110 grocery stores, announced in 1999 that it would no longer sell North Atlantic swordfish, marlin, "orange roughy", or "Chilean sea bass" because these species are endangered because of overfishing. Paul Gingerich, Meat and Seafood Purchasing Director, commented, "We need to be proactive to save these species for future generations." Whole Foods Market a national chain of over 170 stores (out of over 225,000 U.S. retail food stores), in 1999 became the first U.S. retailer offering Marine Stewardship Council (MSC)–certified seafood. The MSC "is an independent, global, non-profit organisation whose role is to recognise, via a certification programme, well-managed fisheries and to harness consumer preference for seafood products bearing the MSC label of approval."[4]

QUESTION 18-17. When you buy seafood or eat seafood at a restaurant, have any of your choices been labeled as "MSC-certified" or "sustainable"? Do you think most consumers would preferentially buy or eat seafood so marked? Explain your answer.

Recreational Fishing

What percentage of total fish landings in the United States come from recreational fishing? If you guessed a few percentage, a common belief, you would be wrong. A study published in *Science* magazine[5] reported that recreational catches account for nearly a quarter of the total take of overfished populations like red snapper. According to lead author Felicia Coleman, "The conventional wisdom is that recreational fishing is a small proportion of the total take, so it is largely overlooked. But if you remove the fish caught and used for fish sticks and fishmeal (pollock and menhaden)—two strictly commercially caught species that account for over half of all U.S. landings - the recreational take rises to 10% nationally."[6]

Co-author Will Figueira added this:[7]

"With over ten million saltwater recreational anglers in the U.S., and recreational fishing activity growing as much as 20% in the last 10 years, their aggregate impact is far from benign. Recreational anglers are operating below the radar screen of management. While the individual may take relatively few fish, we show that a few fish per person times millions of fishermen can have an enormous impact."

---

4. http://www.msc.org.

5. Coleman, F. C., W. F. Figueira, J. S. Ueland, and L. B. Crowder. 2004. The impact of U.S. recreational fisheries on marine fish populations. *Science* 305:1958–1960.

6. As cited in http://www.sciencedaily.com/releases/2004/08/040827094111.htm.

7. Op. Cit.

QUESTION 18-18. Global fisheries has been called a modern *Tragedy of the Commons*, referring to the situation in which there is open access to limited resources (fish stocks). Stake holders (in this case fishers), believing that others will use the resource if they do not maximize their own use invariably deplete the resource. Discuss whether you think that partitioning the oceans into privately held sections would solve this problem.

Media Analysis 1

QUESTION 18-19. Make a list of the seafood you have eaten in the past week or two. Then using the information at the website http://www.mbayaq.org/cr/seafoodwatch.asp, fill in Table I 18-3 here or on our website (http://www.jbpub.com/abel/).

QUESTION 18-20. Based on your seafood chart, analyze whether your diet was seafood–friendly and was a part of sustainable fisheries or whether it contained seafood whose capture or culture had an adverse environmental impact.

Media Analysis 2

Go to www.npr.org, and search for "Commercial Fleets Killing Off Giant Fish."[8] Listen to and evaluate a 4-minute NPR report on how commercial fishing is killing off the ocean's biggest predatory fishes.

QUESTION 18-21. What is the purpose of the research biologist Ransom Myers of Dalhousie University is conducting?

QUESTION 18-22. What are his findings?

TABLE I 18-3	Table for Question 18–19. Labeled "Seafood You've Recently Eaten"	
Seafood Item	**Where it's from**	**Conservation Notes**

8. Or go directly to http://www.npr.org/templates/story/story.php?storyId=1263623.

QUESTION 18-23. Why does Myers say that removing the big fish is a threat to the survival of entire species?

QUESTION 18-24. Why are problems compounded by the longline fleet?

QUESTION 18-25. What is Myers' "one solution" to the problem?

QUESTION 18-26. Identify some of the consequences of the extinction of large pelagic fishes.

QUESTION 18-27. Why do fishers and government scientists find Myers' plan unrealistic? Explain whether you think *all* fishers and government scientists find Myers' plan unrealistic.

QUESTION 18-28. What is your reaction to Mike Sissenwine's statement that Myers' plan is "a social and economic choice that society makes, not scientists?"

QUESTION 18-29. Discuss whether you think Sissenwine's statement that this issue is "not about luxury white table cloth products but is about survival" is fair and accurate.

QUESTION 18-30. John Sackton, president of Seafood.com, argues that sustaining a fishery can be done even if the larger adults are lost, as long as the fish get a chance to reproduce before they are caught. Identify the strengths and weaknesses of his argument.

QUESTION 18-31. What approaches would you take to ensure recovery and survival of the large predatory fish discussed in this report?

Bycatch or Bykill? Dolphin-Safe Tuna and Turtle-Safe Shrimp[1]

KEY QUESTIONS

What is bycatch?

Should "bycatch" be called "bykill"?

How serious is the bycatch issue?

What do consumers think "dolphin safe" means?

What is the exact meaning of "dolphin safe"?

What are the social, economic, and environmental costs and benefits of dolphin-safe fishing methods?

Can we identify an ecologically acceptable impact for commercial fishing?

Are there really dolphin-safe tuna, turtle-safe shrimp, and seabird-safe fish?

Our ship Esperanza has been monitoring UK fisheries for evidence of dolphin deaths in trawler nets. Yesterday we found what we had hoped not to: five dead dolphins, floating in the vicinity of two sets of pair trawlers.

—Dolphin "bycatch" death evidence, *Greenpeace* (February 7, 2004)[2]

One of the main problems with large-scale "harvesting" of wild marine organisms is that most commercial fishing techniques are indiscriminate; that is, they cannot selectively capture only the target species. As a result, as much as 25% of the total global commercial catch is wasted or unused, and its removal from living systems has ecological repercussions (see Issue 19). This catch is known as *bycatch* and refers to undersized fish (based on size limit regulations), fish of low commercial

---

1. Robert F. Young of the Marine Science Department of Coastal Carolina University contributed to this Issue.
2. Available at http://www.greenpeace.org/international/news/dolphin-bycatch-death-eviden.

value, and other nontarget animals. These may include benthic (bottom-dwelling) organisms such as sponges, worms, sea stars, crabs; and rays, pelagic organisms such as sharks (see Issue 13), dolphins, whales, and sea turtles; and even seabirds (Figure I 19-1). Bycatch is also known as *bykill* because although some bycatch is returned to the water and may survive being caught, most dies in nets or on longlines or is returned to the water dead or dying.

Among the most harmful of all fishing activities is trawling for groundfish and shrimp (see Issue 20). Trawling (See Figures I 21-1a and I 21-b) has been compared with clearcutting a forest. In addition to damaging the ocean bottom, as much as 90% of the trawl contents may be nontarget and hence unused species, sometimes called trash fish by fishers and at times including endangered sea turtles (Figure I 19-2a and I 19-2b). Thousands of shrimp boats ply U.S. waters. Additionally, the U.S. imports wild-caught shrimp from nearly 40 countries. Although shrimp trawling may cause

 FIGURE I 19-1 A dead bird entangled in fishing gear.

(A)

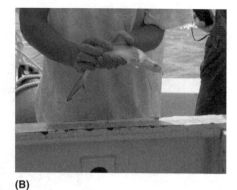

(B)

FIGURE I 19-2A/B   Examples of bycatch. (A) Sea turtle, before *turtle excluder devices (TEDs)* reduced mortality (photo courtesy of the National Oceanic and Atmospheric Administration/ Department of Commerce). (B) Newborn Atlantic sharpnose shark caught in a trawl (and released alive). [Photo by D. C. Abel.]

extreme environmental damage, many consumers are virtually unaware of the dimensions of its destructiveness. A better publicized and more galvanizing bycatch issue is the capture of dolphins by tuna fishers. This is the major focus of this Issue.

Dolphins and Tuna in the Eastern Tropical Pacific

The eastern tropical Pacific Ocean (ETP), an area of approximately 8 million square miles, is one of the world's richest sources of commercially important tunas. The ETP fishery for yellowfin tuna (*Thunnus albacares*) has been called one of the most important fisheries in the world. Yellowfin and skipjack tunas (*Katsuwonus pelamis*) are mainstays of the canned light meat tuna industry. The ETP fishery for albacore (*Thunnus alalunga*), whose flesh is the basis of the white-meat tuna industry, is small by comparison.

Two methods have been used to catch yellowfin and skipjack tunas in large-scale fisheries in the ETP. In pole-and-line fishing, a technique no longer practiced in the ETP, rugged commercial fishers used stout rods to catch tunas, which frequently bit unbaited hooks during their feeding frenzy. Worldwide, according to the Food and Agricultural Organization of the United Nations, 14% by weight of the world's commercial tuna are caught on pole and line.

A more productive method of catching yellowfin, skipjack, and other tunas is purse seining. Globally, about 60% of the world's commercial tuna are caught in purse seines. Pelagic longlining, in which multiple hooks are set at intervals along a buoyed horizontal line stretching for miles, accounts for 14% of world commercial tuna catch, mostly albacore. Gillnets and trolling account for the remaining tuna catch.

In purse seining (Figure I 19-3), a school of fish is encircled by setboats using a net that may be 2 km long and 200 meters deep. A purse line attached to the bottom of the net is then pulled in to the purse seiner, trapping the tunas and other organisms

FIGURE I 19-3 Purse seining. What does the net look like below the surface? [Photo courtesy of the National Oceanic and Atmospheric Administration/Department of Commerce.]

unfortunate enough to be in the same location. Vessels from about 12 nations, including the United States, purse seine in the ETP for tuna.

Tunas frequently congregate around floating objects, such as tree trunks, rafts of seaweed, or other objects, as do two kinds of dolphins, northern offshore spotted (*Stenella attenuata*) and eastern spinner (*Stenella longirostris*) dolphins. Surprisingly, enough such objects enter the ocean to be worthwhile to commercial fishers. The association between tunas and dolphins is thought to benefit the tunas, which can easily follow dolphins and take advantage of the latter's superior prey-finding abilities. Setting nets around dolphins typically catches the largest tunas and is thus is considered the more desirable method by commercial fishers.

Helicopters or spotter plains locate aggregations of tunas and dolphins because dolphins are noisy and disturb the sea surface, and they report coordinates to the purse seiner. The netting process, which can take 2 to 3 hours, does not discriminate between the tunas and the dolphins, which stay together throughout the process (Figure I 19-4). Dolphins can be killed during the process by entanglement and drowning. More may die later because of the delayed effects of severe trauma caused by the stress of the long capture process. Experts estimated that the purse-seine fishery for tuna killed more than 1.3 million eastern spotted dolphins in the ETP between 1959 and 1990.

Policies to Curb Dolphin Mortality

National and international legislative attempts to curb the killing of dolphins during tuna seining include an agreement reached by the Inter-American Tropical Tuna Commission (IATTC) in 1976. This program sought (1) to determine dolphin mortality, (2) to reduce it such that dolphin populations were not threatened and accidental killing was avoided, and (3) to maintain a high level of tuna production. The chief result of this effort was

FIGURE I 19-4 Trawl with entrapped dolphins. Estimate the number of dolphins in the net. How has the number of entrapped dolphins in tuna fisheries changed over time? [Photo courtesy of the National Oceanic and Atmospheric Administration/Department of Commerce.]

the placement of observers on one third of all vessels fishing in the ETP. As a result, the first reliable estimates of dolphin mortality were made.

This was followed by a series of policies designed to lower dolphin mortality, including new provisions of the Marine Mammal Protection Act in 1988 and 1990, the Dolphin Protection Consumer Information Act of 1990, the Agreement for the Conservation of Dolphins of 1992 (more commonly called the La Jolla Agreement), the Panama Declaration of 1995, and the International Dolphin Conservation Program Act of 1999.

These resulted in 100% observer coverage of ETP tuna seiners and established international limits of fewer than 5000 dolphins killed by 1999. Moreover, criteria were instituted for labeling canned tuna as "dolphin safe."

QUESTION 19-1. What do you think a typical American consumer thinks after reading the label "dolphin safe"?

The commercial fishing industry also instituted its own dolphin conservation measures that reduced dolphin bycatch by as much as 99% in the ETP. The first was a procedure known as the "backdown" in which the purse-seining vessel essentially moves in reverse, pushing the net in such a way as to force a portion of the purse seine to sink, allowing dolphins to escape easily. A second measure involved changing the size of the mesh in a portion of the net so that dolphins no longer became entangled.

Dolphin mortality has decreased in the ETP as a result of conservation measures; however, the issue remains controversial, and repercussions have been felt ecologically, economically, socially, and politically, as we discuss later here.

· ·
Dolphin Mortality 1987–2002

In this analysis, you graph data on dolphin mortality from the IATTC and calculate the percentage decrease from 1987 to 2002. Then you compare the rate of mortality of dolphins from purse seining to their population growth rate. Finally, you examine alternative methods of catching tuna and assess the environmental and social impact of these.

- ## Have Dolphin Populations in the ETP Been Threatened by Incidental Capture in Tuna Purse Seines? Are They Now?

Millions of dolphins have likely been killed in the ETP since the inception of dolphin encirclement as a means of capturing tuna. Today, that number has decreased significantly, although there is disagreement over the magnitude of the decrease. Controversy still exists relative to delayed mortality and sub-lethal effects (e.g., lower reproductive rates) associated with capture stress.

According to Earth Island Institute,[3] an environmental advocacy group,

Federal scientists have determined that dolphin populations in the ETP are not recovering as expected, even with the dramatically lower reported kills of recent years. Harassment of dolphins by tuna fishermen and problems arising from the consequent physiological stress (some dolphin schools are chased and netted as often as three times in one day) are likely factors which cause harm to dolphin health and reproduction. Many dolphins suffer injuries in the nets and die after release, but are not counted by the on-board observer. Mothers are separated from calves, and undercounting may be occurring onboard some Mexican tunaboats.

To determine whether a level of dolphin mortality threatens the stability of their populations, scientists must examine, among other data, the *recruitment rate* of the dolphin population. The recruitment rate is an estimate of the rate at which recently born individuals survive to enter the adult population. In this case, knowing the recruitment rate provides policy makers with an estimate of the dolphin mortality that may be "acceptable," at least from the perspective of population stability. With respect to ETP dolphins, the question is whether mortalities caused by incidental catch in purse seines exceed the recruitment rate. We examine this issue more fully later here.

- ## Do Alternative Methods Reduce Dolphin Bycatch?

Three methods of purse-seining in the ETP are most common: school sets involve nets set around solitary schools of tuna; dolphin sets use nets set around schools of dolphins; and log sets target floating objects such as trees trunks, under which fish congregate. Log sets may also include FAD sets (*Fish Aggregating Devices*), placement of stationary or drifting man-made structures to attract fish.

---

3. http://www.earthisland.org/news/news_immp13.html.

Accurate estimates of fishery-induced dolphin mortality are difficult to make. One estimate is that dolphin sets typically kill as many as 29 dolphins per 1,000 tons of tuna landed. According to the National Oceanic and Atmospheric Administration Southwest Fisheries Science Center, this equals about 1,000 dolphins per year killed in the ETP tuna fishery, mostly from dolphin sets. Dolphin mortality from log and FAD sets is thought to be negligible.

Log and FAD sets thus may reduce dolphin mortality, but they do so at the cost of much increased bycatch of other marine organisms. In our analysis later here, we provide the data to analyze the magnitude and consequences of differences in bycatch between the different methods.

According to the National Marine Fisheries Service, in 1975, dolphin mortality associated with purse seining was about 200,000. The IATTC reported that about 100,000 dolphins were killed in 1989. In 2006, fewer than 900 dolphins were killed. Moreover, about 94% of all dolphin sets occurred with no mortality or serious injury to the dolphins, according to the IATTC.

QUESTION 19-2. What was the percentage decrease in dolphin mortality over the 14-year period from 1975 to 1989? The 17-year period from 1989 to 2006?

QUESTION 19-3. Estimate the average *annual* decrease in mortality over that period?

Divide your answers to Question 19-2 by the number of years (i.e., 14 or 17), or for a more accurate answer, use the formula $k = (1/t)\ln(N/No)$.

The estimated population for all dolphins in the ETP in 1986 was 9,576,000. Incidental kill by the purse-seine fishery in that year was estimated at 133,174.[4] The 2006 ETP dolphin population was approximately 9,000,000, and about 900 were killed.

QUESTION 19-4. What percentage of the estimated dolphin population was killed by tuna seiners in 1986? In 2006?

For dolphins in the ETP, the recruitment rate has been estimated to be approximately 2% of the total population.[5]

QUESTION 19-5. Based on an annual recruitment rate of 2% and 1986 incidental mortality you just calculated for the ETP, do you think dolphin populations were threatened by purse seining in 1986? Show any calculations and explain you reasoning. Also, list assumptions you made in arriving at your answer.

QUESTION 19-6. Based on an annual recruitment rate of 2% and 2006 incidental mortality you just calculated for the ETP, do you think dolphin populations are

4. Wade, P. R., and T. Gerrodette. 1993. Estimates of cetacean abundance and distribution in the eastern tropical Pacific. *Rep Int Whal Commn* 43:477–493.

5. Smith, T. D. 1983. Changes in the size of three dolphin (Stenella spp) populations in the eastern tropical Pacific. *Fish Bull US* 81:1–13.

currently threatened by purse seining? Show any calculations and explain your reasoning. Also, list assumptions you made in arriving at your answer.

QUESTION 19-7. In light of your answer to Questions 19-5 and 19-6, are efforts to reduce dolphin mortality justified from the perspective of threatening the health of the population? Explain your reasoning.

In spite of the apparent success of dolphin-safe methods to reduce dolphin mortality in the ETP, stocks of at least two species, spotted and spinner dolphins, as of 2007 had not recovered. According to the National Marine Fisheries Service, by 2002, populations of northeastern spotted and eastern spinner dolphins were approximately 19% and 29%, respectively, of prefishery levels, and these have not significantly increased since then. One possible explanation is that the stress of being chased by speedboats, encircled, and captured may result in a delayed mortality or other acute effects not leading to death, such as altered reproductive output, increased vulnerability to predation, etc. Also, separation of mothers and calves may result in decreased survival of the latter. Alternatively, perhaps additionally, other changes in the ETP ecosystem (e.g., declines in abundance of plankton) may have led to declines in dolphin populations.

QUESTION 19-8. Explain whether this new information on dolphin populations causes you to change your answer to Question 19-7.

Recall the two common methods of purse seining, dolphin sets, in which nets are dropped around schools of dolphins, and log sets in which nets encircle floating objects such as trees and FADs, under which fish congregate. A major difficulty with making tuna fishing dolphin-safe concerns the size of tuna captured by these two methods.

Dolphin sets may kill 29 dolphins per 1,000 tons of tuna. Log sets kill less than one. Dolphin sets catch large yellowfin tuna. Log sets catch skipjack, bigeye, and small yellowfin tuna.

Modal lengths and weights of the tuna caught during log sets for 1994 were 47.5 cm and 2.1 kg for skipjack tuna. For dolphin sets, the distribution was bimodal, 103 and 138 cm and 23 and 57 kg.

QUESTION 19-9. Identify a possible disadvantage to tunas of capturing such small fish using log sets. Explain your answer.

In addition to capturing immature tuna, log sets also create another problem: bycatch. In this case, during log sets, bycatch can include mahi mahi, sharks, wahoo, rainbow runners, other small fish, billfish, yellowtail, other large fish, triggerfish, and sea turtles.

A multiyear analysis of 10,000 dolphin sets showed that 5,340 dolphins were captured, along with 1.56 million small tunas, 11,046 sharks, 98 sea turtles, and 3,641 "other small fish." The respective numbers for 10,000 log sets were 36 dolphins, 103.2 million small tunas, 140,185 sharks, 456 sea turtles, and 264,886 other small fish.

QUESTION 19-10. For each type of previously mentioned bycatch, calculate the ratio of organisms caught during log sets to those captured during dolphin sets (e.g., 36/5,340 dolphins), and complete this table.

Ratio	Log Set	Dolphin Set
Dolphins 1:148.3	36	5,340
Tunas (small)		
Sharks		
Sea Turtles		
Other		

QUESTION 19-11. For each type of bycatch, in both dolphin and log sets, calculate how many small tunas etc. were captured for each dolphin etc. Fill in the table.

	Log Set	Dolphin Set
Tuna per dolphin	2.87×10^6	292
Sharks per dolphin		
Sea turtles per dolphin		
Other per dolphin		
Tuna per shark		
Tuna per turtle		

QUESTION 19-12. Which method—log or dolphin sets—do you think is more ecologically sound? Explain your answer.

QUESTION 19-13. Like most issues, the dolphin–tuna controversy has many dimensions. In Mexico, as many as 15,000 jobs in the tuna fishing and canning industry have been lost, and this loss has been attributed directly to the "dolphin-safe" issue. In 1991, an embargo was placed on Mexican tuna as a result of Mexico's tuna fleet's killing too many dolphins. Do you think an embargo was justified in light of the number of jobs lost in a relatively poor country? Do you think regulations should be relaxed? Justify your answer.

In the United States as a result of a 2004 legal decision, all canned tuna must have been caught on a trip on which no dolphins were chased, encircled or killed, which is the original meaning of "dolphin safe."

QUESTION 19-14. Conservation measures reduced dolphin mortality from 15 per set in 1986 to 3.1 in 1991 and even fewer today. Should the current definition of "dolphin safe" be expanded to include methods of dolphin sets that reduce dolphin mortality? Justify your answer.

QUESTION 19-15. Dolphins are relatively intelligent animals. What role does this play in your assessment of the issue? How do you weigh killing intelligent animals versus endangering stocks of sea turtles or sharks?

Turtle-Safe Shrimp

In 2000, the Associated Press reported this:

Since mid-April, about 280 dead sea turtles, mostly threatened loggerheads, washed up on ocean beaches on Ocracoke and Hatteras islands. Gear from large-mesh gill nets was found on four turtle bodies. The turtles were killed in greater numbers in a week's time than the average on all state beaches in a year.

This particular kill was attributed to commercial fishers who fished North Carolina's waters in March and April 2000 for monkfish. Sea turtle kills caused by commercial shrimping are even more common; in the early 1990s, according to conservationists, shrimp trawlers in Texas killed more than 11,000 sea turtles annually.

The use of turtle excluder devices (TEDs) on trawl nets and circle hooks (see Figure I 20-2) on longlines can significantly reduce sea turtle mortality. TEDs can be retrofitted onto existing nets. They work by shunting large, heavy objects out of the net through a trap door while allowing shrimp to pass to the back of the net. Sea turtle bycatch was reduced an estimated 97% by the use of TEDs. In two studies, circle hooks resulted in significantly lower turtle catch compared with traditional "J"-style hooks. In one case, leatherback and loggerhead turtle catch was reduced by 65% and 90%. The circle hooks replacing the J-hooks are slightly larger. Turtles apparently feel the hook and do not bite down and thus avoid being hooked.

Sea turtles also die as a result of dredging associated with mining sand in offshore waters for beach nourishment. Diamondback terrapins, found in coastal estuaries along the U.S. Atlantic and Gulf coasts, may be threatened as well as a result of crab pot mortality, road kills, impacts with boats and personal watercraft, and habitat loss.

QUESTION 19-16. Discuss whether you think consumers (including you) would change your seafood eating habits if you knew that dolphins, sharks, turtles, seabirds, and other marine life were killed during capture of your meal.

Media Analysis 1

Go to www.loe.org (*Living on Earth*), and search for "Bypassing Bycatch."[6] Scroll down to the August 22, 2003, story, and listen to the 15-minute report. Then answer the following questions.

---

6. Or go directly to www.loe.org/shows/shows.htm?programID=03-P13-00034.

QUESTION 19-17. Solomon-Greenbaum refers to bycatch as trash fish, a term used by commercial fishers. Loggers sometimes refer to unwanted trees in a clearcut as trash trees. Why do you think fishers and loggers call these organisms trash? Do you think such terminology is appropriate? Why or why not?

QUESTION 19-18. If Joe and Vin net more than 400 pounds of cod (the legal limit) while fishing for flounder, they must toss the excess dead and dying cod overboard. What are the costs and benefits of this regulation? Does it make sense to throw back dead or dying organisms that otherwise could be consumed by humans? Explain your answer.

QUESTION 19-19. Explain how Joe's experimental net differs in both structure and function from a typical net.

QUESTION 19-20. Characterize Dave Marciano's opinion about modifying fishing gear to reduce bycatch. Is his rationale logical? Explain.

QUESTION 19-21. What does Dr. Sissenwine of the Northeast Fisheries Science Center say the "main problem" is? Explain what he means.

QUESTION 19-22. Which term, bycatch or bykill, is most accurate? Explain which term should be adopted by scientists, fishery managers, and regulators.

Media Analysis 2

Go to www.npr.org, and search for "U.S. Fishing Fleet Among World's Most Wasteful."[7]

Listen to the 4-minute report from 2005, and then answer the following question.
QUESTION 19-23. Summarize the issue of bycatch as presented in the report. Include an explanation of the findings of the published study referred to in the report, the different points-of-view, as well as possible solutions.

---

7. Or go directly to www.npr.org/templates/story/story.php?storyId=5035061.

Tragedy of the Trawl: Clearcutting the Ocean Floor

What is trawling?

How severe is the damage caused by trawling?

How widespread is this damage?

Which marine ecosystems are threatened by trawling?

Are there ways to minimize the damage?

Who is responsible for the damage?

Bottom trawling is like fishing with bulldozers. It's devastatingly efficient in one sense; it's a way to get fish relatively easily and painlessly—if you don't mind killing all of the life on the bottom to catch them.

—Elliot Norse, president of the Marine Conservation Biology Institute

Bottom Trawling Is Like Fishing With Bulldozers

Do you eat seafood? According to the National Oceanic and Atmospheric Administration, Americans consumed 16.5 pounds of fish and shellfish per person in 2006 (4.9 billion pounds in all), including 3.9 pounds of canned fish and 4.4 pounds of shrimp. World per capita consumption in 2003 was 35.5 pounds.

Are you aware of how the seafood on your plate was caught or of the environmental impact of the method of capture? Probably not. Among the most common methods of catching seafood are trawling, purse seining (see Issue 19), dredging, and longlining. Other more destructive methods such as dynamiting and poisoning still are used in some areas. In this Issue, we consider trawling.

Trawling

Trawling is typically used for shrimp and other bottom-associated species like Atlantic cod and Atlantic scallops. In this method, a 10 to 130 m long net is dragged across large areas of the bottom (Figure I 20-1a and I 20-1b), collecting or destroying virtually everything in its path, including endangered sea turtles, deep-sea corals, and many other nontargeted species. Trawling, which became popular with the advent of the diesel engine in the 1920s, is practiced worldwide on virtually every different bottom type.

A large percentage of fish caught in trawls is considered bycatch, that is, species that are not the ones targeted for capture. This includes adults of many species that are not popular in the market, juveniles too small to retain because of strict size limits on fishing, or nonfood species such as turtles, seabirds, and dolphins.

Bycatch is also known as bykill since must bycatch die in nets or on longlines, or are returned to the water dead or dying. As much as 90% of the trawl contents may be non-target and hence unused species, sometimes called trash fish by fishers (see Issue 21).

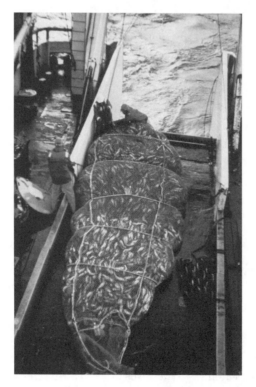

⭐ FIGURE I 20-1A 16,000 pound catch of pollock coming aboard the trawler MILLER FREEMAN. [Photo courtesy of NOAA, U.S. Dept,. of Commerce.]

FIGURE I 20-1B Picture of small shrimp trawl being deployed from a Coastal Carolina University research vessel. [Photo by D. C. Abel.]

Innovations such as turtle excluder devices (Figure I 20-2) shunt large objects like sea turtles out of the net and thus may reduce bycatch. The use of turtle excluder devices, however, is neither universal nor entirely successful.

Shrimp trawling is widespread and may cause extreme environmental damage. Most consumers, however, are unaware of the destructiveness of this fishing method. Thousands of boats ply U.S. and international waters. Saturation trawling, in which an area is repeatedly fished until virtually no fish or shrimp are left, has been compared with clearcutting a forest. Complete recovery in both cases may take centuries.

The comparison between trawling and forest clearcutting breaks down when one considers the area involved. Approximately 100,000 km² of forest are clearcut annually, whereas an area 150 times as large is trawled! This is an area roughly equal to the area of the lower 48 U.S. states and Canada combined.

Media Analysis 1

Go to www.nap.edu (The National Academies Press), and search for "Effects of Trawling and Dredging on Seafloor Habitat."[1] Questions 20-1 to 20-5 refer you to sections of the online text (which you can access without cost).

---

1. Or go directly to www.nap.edu/catalog.php?record_id=10323.

FIGURE I 20-2 Turtle Excluder Device (TED). Turtles and other large objects are deflected through a trap door (not shown) by the oval ring and bars. [Photo courtesy of NOAA/U.S. Department.]

QUESTION 20-1. Go to page 20 of the report and discuss the effect of trawling on Florida's Oculina Banks.

QUESTION 20-2. Go to page 65 of the report for Questions 20-2 through 20-4. What is the most challenging aspect of evaluating the effect of trawling and dredging on marine habitats?

QUESTION 20-3. What communities are most vulnerable to acute and chronic physical disturbance?

QUESTION 20-4. In what parts of the United States are trawling and dredging heaviest?

QUESTION 20-5. Identify which of the three management tools you think would protect marine ecosystems the best. Explain your answer (see page 66 of the report).

QUESTION 20-6. Identify the stakeholders involved in the issue of trawling. Then, for each stakeholder, identify their point of view on the issue. Which stakeholder should have the most influence? Explain your answer.

QUESTION 20-7. Commercial fishers argue that regulation of trawling should be delayed until rigorous scientific data documenting any harmful effects of trawling on the environment are collected. Assess the costs and benefits of waiting, and then discuss whether you think regulation should be enacted immediately or be delayed. Recall the Precautionary Principle. Should it be applied here? Why or why not?

QUESTION 20-8. Managing trawling impact has social, economic, and environmental consequences. How much should each of these be weighed in deciding which tools to use? Explain your answer.

Media Analysis 2

Go to www.npr.org, and search for "Commercial Fishing."[2] Listen to the 4.5-minute report from 2002, and then answer the following questions.

QUESTION 20-9. What were the "strict new regulations" that commercial fishers were required to follow?

QUESTION 20-10. According to the report, some are concerned that these regulations might collapse the region's fragile fishing industry. Would you support the regulations if collapsing the regional commercial fishing industry is a likely outcome? A possible outcome? Explain you answer.

QUESTION 20-11. Commercial fishing entails a financial risk (and is heavily subsidized), even in the absence of regulation. Explain whether you would support (a) increased prices for seafood and/or (b) a tax on seafood if either or both of these would enable commercial fishers to survive financially.

QUESTION 20-12. Explain whether you agree or disagree with the statement that all fish need to be managed at a sustainable level.

QUESTION 20-13. Give some examples of the cultural impacts of commercial fishing. Should cultural impacts of fishery management decisions be considered along with the environmental impacts? Explain.

QUESTION 20-14. Some have argued that regulating commercial fishing is putting fish before people. Do you agree? Justify your answer using principles of critical thinking.

---

2. Or go directly to www.npr.org/templates/story/story.php?storyId=1143369.

Can Aquaculture Replace Capture Fisheries?

KEY QUESTIONS

What is aquaculture?

What are the environmental impacts of aquaculture?

How fast is aquaculture growing?

Where is it growing the fastest?

As of 2004, the U.N.'s Food and Agriculture Organization estimated that there were 4 million fishing boats of all types plying the world's oceans, with 1.3 million of them being "decked" vessels, almost all of them powered. This world fishing fleet has been extraordinarily successful, such that "capture" fisheries are nearly fully exploited and in many cases are severely overexploited. Indeed, capture fishery harvest levels have plateaued and show signs of declining; however, total fish production is increasing as a result of aquaculture, mainly in China. The Chinese have been masters of aquaculture for centuries if not millennia. Today, Chinese farmers routinely integrate rice cultivation with fish farming, growing several species of carp in rice paddies. The advantages are clear: Carp species are filter feeders or herbivores; in either case, they do well in ponds fertilized with manure. The manure feeds grasses, which can be fed on by herbivorous carp, and the grasses provide habitat where phytoplankton and zooplankton live, which in turn are food for filter-feeding carp. Thus, this style of aquaculture may reduce stress on already threatened marine environments. In contrast, several countries (e.g., Norway, Chile, the United Kingdom, and the United States) are beginning to grow carnivorous species like salmon, which actually put additional stress on wild fish stocks, as 3–5 kilograms of wild-caught fish must be fed to salmon to produce a kilogram of meat.

Other concerns arise out of large-scale salmon farming: waste-production, the deterioration of the wild gene stock, diminished nutritional value of farmed salmon, fish captured for salmon feed reduces the availability of forage species in the wild, and changes to native habitat. Salmon kept in pens and fed large quantities of fish meal produce in turn large quantities of concentrated waste: In Norwegian salmon farms, for example, the fish produce as much waste as Norway's 4 million humans. Additionally, farmed salmon are bred for rapid growth and not to develop traits that will aid their survival in the wild. Salmon that escape their pens can and do breed with wild salmon, however, and could in time reduce their wild counterparts' ability to survive in the open ocean. There is some evidence that farmed salmon are not as nutritious as wild-caught fish, deficient in important substances such as omega-3 fatty acids. Forage fish, small, schooling fish like herring, anchovies, smelt, and sardines, are important food sources for larger fish. They are pivotal components of marine ecosystems and are subjected to heavy fishing pressure. No current policies effectively protect forage fish species. In Chile's Patagonian lakes, salmon farming has doubled in the last 10 years. Nutrient pollution, invasive species (salmon farmed there are predominantly Atlantic salmon), the risk of disease, and the introduction of toxic chemicals all threaten this system.

Shrimp mariculture is growing rapidly as well (Figure I 21-1). Tiger shrimp, the main variety grown, is a tropical crustacean; thus, areas developed for shrimp farms tend to be areas in the coastal tropical ocean, itself under severe strain from the growth of human populations (see Issue 2).

Large-scale marine aquaculture requires drastic land-use changes. Shrimp farming is stressful in at least two ways. First, areas devoted to shrimp farming tend to be cleared mangrove swamps (Figure I 21-2). Mangroves (salt-tolerant trees) are of critical

FIGURE I 21-1 Shrimp farm.

⭐ FIGURE I 21-2 A mangrove swamp cleared for the shrimp farm shown in Figure I 21-1. How will the susceptibility of the coast to tropical storm damage change in this area as a result of mangrove clearance?

importance as nurseries for juvenile fish and other marine animals, and the mangrove roots trap and bind sediment washing offshore from rivers. This sediment, if left unchecked, can smother offshore reefs. Reefs, in turn, are important fishing grounds for many subsistence fishers, especially in Southeast Asia.

Mangroves also provide important protection to the coast from the impacts of tropical storms. Ironically, the mangroves cleared for shrimp farms are more productive biologically than the farms that replace them, in that the mangroves could supply a greater harvest of fish. For example, the impact that the loss of the mangrove ecosystem has on local Ecuadorian fishermen and their families is illustrated by a study concluding that the loss of one acre of mangroves will result in a drop in the harvest of wild shrimp and fish by 676 pounds per year.[1]

Second, shrimp are fed fish meal, which, like salmon, comes from wild stocks.

Some Advantages of Aquaculture

Aquaculture, if carried out in a manner sensitive to land-use conservation and the avoidance of additional stresses on marine fish stocks, could be an important addition to a human food base that is already threatened by land degradation, pollution, and the addition of 80 million new mouths to feed each year.

1. www.earthsummitwatch.org.

The Earth Policy Institute's President Lester Brown is an agricultural expert. Here is what he said recently about the advantages of fish farming.[2]

Cattle require some 7 kilograms of grain to add 1 kilogram of live weight, whereas some farmed fish require less than 2 kilograms of feed to produce one kilogram of live weight. Water scarcity is also a matter of concern since it takes 1,000 tons of water to produce 1 ton of grain, but the fish farming advantage in the efficiency of grain conversion translates into a comparable advantage in water efficiency as well, even when the relatively small amount of water for fishponds is included.

Finally, large-scale shellfish farming can provide high-quality food and help clean polluted water at the same time, as mussels and clams are filter feeders and can filter large quantities of algae from the water daily.

The Growth of Aquaculture

From 1984 to 2005, world aquaculture production rose from 6.9 to 47.8 million metric tons, whereas in China it increased from 2.2 to 30.6 million metric tons to 2004.[3] Fish farming may soon overtake cattle ranching as a source of protein if these trends continue.

QUESTION 21-1. Using the data in Table I 21-1, on the axes below, or on our website (http://www.jbpub.com/abel/), plot the growth of aquaculture and wild fish catch ("Total Catch") over the period of 1950 to 2005.

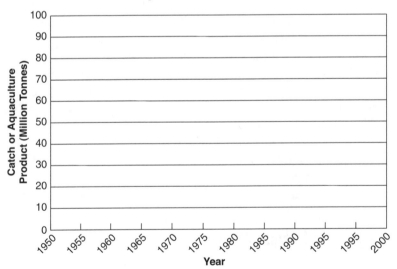

QUESTION 21-2. Describe any trends apparent from your graphs, for example, which source is growing or changing fastest? Have there been intervals when rates of change differed from others? List them.

---

2. Worldwatch Institute. 1998. Vital Signs. New York: W.W. Norton and Co.
3. www.worldwatch.org/chairman/issue/001003.html.

TABLE I 21-1	World Fish Catch, Aquaculture, and Total Harvest 1950-2005*	
Year	Total Wild Catch (million tonnes)	Aquaculture (million tonnes)
1950	19.2	1.5
1951	21.5	1.7
1952	22.9	1.7
1953	23.7	1.8
1954	25.2	1.9
1955	26.4	2.1
1956	27.8	2.1
1957	28.6	2.3
1958	30.2	2.5
1959	33.4	2.8
1960	36.4	3.0
1961	39.5	3.2
1962	42.9	3.3
1963	44.2	3.4
1964	48.5	3.6
1965	49.0	3.7
1966	52.6	3.9
1967	55.6	4.0
1968	56.5	3.9
1969	57.4	3.7
1970	58.2	3.6
1971	62.4	3.8
1972	58.4	3.8
1973	59.0	3.9
1974	62.6	4.0
1975	62.4	4.1
1976	64.6	4.8
1977	63.4	4.8
1978	65.3	4.8
1979	66.1	5.0
1980	67.0	5.2
1981	69.4	5.4
1982	71.1	5.6
1983	71.6	5.9
1984	77.6	6.7

(*Continues*)

TABLE I 21-1 (*Continued*)		
Year	Total Wild Catch (million tonnes)	Aquaculture (million tonnes)
1985	79.1	7.7
1986	84.6	8.8
1987	85.0	10.1
1988	88.6	11.7
1989	89.3	12.3
1990	85.9	13.1
1991	84.0	13.7
1992	85.0	15.4
1993	86.0	17.8
1994	91.0	20.8
1995	92.0	24.4
1996	93.0	26.8
1997	94.0	28.8
1998	87.6	30.7
1999	93.7	33.4
2000	95.5	35.5
2001	92.8	37.9
2002	93.0	40.4
2003	90.5	42.7
2004	95.0	45.5
2005	93.8	47.8

*Compiled by Worldwatch Institute (http://www.worldwatch.org) from U.N. Food and Agriculture Organization Yearbooks.

"Finfish" such as tilapia, salmon, carp, and flounder make up about one half of aquaculture production. Molluscs, mainly oysters and mussels, account for one fourth. Shrimp and prawns (i.e., crustaceans) make up the final fourth. Two thirds of aquaculture takes place in inland rivers, lakes, ponds, and artificial tanks. Coastal marine aquaculture and mariculture account for the remainder.

Aquaculture output, like capture-fisheries production, is used for both fishmeal and canned, frozen, and fresh products. The growth rates for aquaculture production are remarkable. Shrimp mariculture alone grew by 350% between 1984 and 1993, in response to both market demands and often to government subsidies.[4]

---

4. Malcolm C., M. Beveridge, L. G. Ross, and L. A. Kelly. 1994. Aquaculture and biodiversity. Ambio, December.

Fish Farming In China

China has long led the world in total production from fish farming, and produced 70% of the world's farmed fish in 2007, valued at more than $9 billion. Chinese exports are beginning to dominate the U.S. market. For example, in 2007 imports of Chinese catfish reached 6,600 metric tonnes and made up 49% of catfish imports. However, in 2007, the chemical melamine was detected in one sample of catfish imported from China. According to the U.S. Food and Drug Administration, in March/April 2007, the agency began studying pet deaths in the United States related to the consumption of pet food contaminated with melamine. Melamine is used in fertilizer in China but has been used to enhance nitrogen levels in feed, presumably in fish feed as well as pet feed.

Meanwhile, imports of Chinese-farmed fish were testing positive for high levels of banned antibiotics (used also in Chilean salmon farming), which caused cancer in laboratory tests. The antibiotics were apparently fed to fish to counter the effects of intense crowding in ponds, as well as polluted water.

Genetically Modified Fish

The company Aqua Bounty has produced a genetically modified (GM) salmon that grows six times faster (but not bigger) than their normal counterparts and reach harvest size in about a year and a half (compared with 3 years). According to the company, the fish are "basically identical" to other salmon, but with a slightly lower fat content. The company has also pledged that only sterile, all-female GM salmon will be marketed, for use exclusively in aquaculture.

Phillip Nabors, president of Mustard Seed Market & Cafe in Akron, Ohio, urged the company to label any transgenic salmon products as genetically-modified. He also wanted the environmental impacts and ethical issues addressed. "Once [the engineering is] done, you can't put the genetic genie back in the bottle," he said.[5]

Elliot Entis, president of Waltham, Massachusetts-based Aqua Bounty, accused Nabors of having a "rigid environmental bias."

QUESTION 21-3. Based on what you have learned in this issue, discuss the advantages and disadvantages of the use of GM salmon. Explain whether you think on balance GM salmon are a good idea.

Media Analysis 1

Go to www.npr.org, and search for "Fish Farming Potential Threat to Wild Salmon."[6] Listen to the first 11 minutes and 15 seconds of the 17-minute program, and then answer the following questions.

---

5. As quoted in Aqua Bounty Farms to begin selling genetically modified salmon eggs. Available at http://www.mindfully. org/GE/GE2/Salmon-Aqua-Bounty.htm
6. Or go directly to http://www.npr.org/templates/story/story.php?storyId=6210154.

QUESTION 21-4. According to the National Academy of Science report, what is infesting wild salmon in British Columbia?

QUESTION 21-5. What is the source of the infestation?

QUESTION 21-6. How is the introduction of farmed salmon into the environment where they can breed with wild salmon harmful to wild salmon?

QUESTION 21-7. Explain a criticism of the NAS paper as described by Dr. Hilborn.

QUESTION 21-8. According to Dr. Hilborn, what are the safest, healthiest salmon to eat?

QUESTION 21-9. What solution did environmental groups propose to the problems caused by raising salmon in pens?

QUESTION 21-10. How did Dr. Hilborn characterize this proposal?

QUESTION 21-11. Should the public have a right to buy fish raised as cheaply as possible without regard to the environmental impacts? Why or why not?

Media Analysis 2

Go to www.npr.org, and search for "Ocean Researchers Take Fish Farming Offshore."[7] Listen to the 7.5-minute program from 2004, and answer the following questions.

QUESTION 21-12. What percentage of seafood eaten in the United States as of 2004 was farm raised, and what was the principal source of the farm raised seafood?

QUESTION 21-13. According to the 2003 *Nature* article referred to in this report, what percentage of "large, valuable fish" had been killed off?

QUESTION 21-14. How did the NPR report characterize the problem of waste from fish farms?

QUESTION 21-15. How many cod were being raised in the submerged 80' × 50' pen?

QUESTION 21-16. What does Environmental Defense spokesperson Becky Goldberg assert is the largest problem facing large-scale farming of predatory fish?

QUESTION 21-17. How did the University of New Hampshire researcher respond to the concern expressed in Question 21-16?

---

7. Or go directly to http://www.npr.org/templates/story/story.php?storyId=3620330.

PART 9

Invasive Species

ISSUE 22

Veined Rapa Whelk in Chesapeake Bay

KEY QUESTIONS

What are exotic (invasive) species?

How are they able to cross oceans?

What enables them to colonize foreign environments?

What threat do they represent to coastal ecosystems?

How can we protect estuaries from nonnative species?

In September 1999, then-U.S. President Bill Clinton issued an executive order that directed the Departments of Agriculture, Interior, and Commerce, the Environmental Protection Agency, and the U.S. Coast Guard to develop an alien species management plan to blunt the economic, ecological, and health impacts of invasive species.

Environmentalists complain that the federal government has been slow to regulate ballast water discharges from freighters—one of the major pathways for exotic aquatic organisms such as the Chinese mitten crab (annual economic cost unknown), green crab (annual economic cost $44 million), and Asian clam (annual economic cost, $1 billion), which are threatening native marine life in San Francisco Bay and as far north as Washington state.

Because the issue involves interstate and international commerce, individual states and counties cannot, under the U.S. Constitution, regulate ballast water. "The West Coast invaders," says Linda Sheehan of the Center for Marine Conservation in San Francisco, "are even burrowing into and weakening flood control levees, which could potentially result in huge losses from property damage during floods."

Illegal Immigration?

Almost daily, rivers, lakes, and seas around the world are being invaded by animals, plants, bacteria, and even viruses from foreign ecosystems. These organisms are called invasive, alien, or exotic species.

FIGURE I 22-1 Ballast water being discharged from a freighter. Describe the kinds of alien species that could be found in this ballast water.

An "invasive species" has been defined by Presidential Executive Order as a species (1) that is not native to the ecosystem in which it is found and (2) whose introduction "causes or is likely to cause economic or environmental harm, or harm to human health."

Although there are many ways to introduce an alien species to a new habitat, the most common means is by "hitchhiking" in the ballast water of ocean-going ships or by attaching to their hulls. Individuals and organizations may also free invasive species by dumping animals and plants (from aquaria, for example) into water bodies. Accidental releases from aquaculture facilities may also contribute, as shown in the following Issue. Figure I 22-1 shows ballast water from a freighter being dumped into a harbor. The United States Geological Survey estimates the annual economic cost of invasive species to exceed $100 billion.

Why Are Invasive Species a Problem?

Imagine a natural ecosystem in which the organisms have evolved and interacted over long intervals of time. For example, the Chesapeake Bay of the Atlantic Coast and the Great Lakes of North America formed more than 6,000 years ago during the last episode of sea-level rise. Predator–prey relationships become well-established. However, in the relative blink of an eye, an alien species can be transported thousands of miles outside its natural range into the presence of other organisms, with which it will immediately begin to interact and possibly outcompete. Unlike native species, an alien species may have no natural predators.

Many if not most alien species do not survive in their new environment, but those that do have one or more of the following attributes:

- They are hardy, perhaps having survived inside a ship's ballast water for thousands of miles.
- They are aggressive and can outcompete native species.
- They are prolific breeders, able to take rapid advantage of any new opportunity.
- They can disperse rapidly, perhaps aided by a planktonic larval stage, allowing the juveniles to be transported far and wide by currents.

In any of these scenarios, invasive species can flourish and reach astonishing numbers, once again aided by the absence of local predators.

Invasive species can inflict damage on ecosystems by

- Outcompeting and driving out native species
- Introducing parasites and/or diseases
- Altering habitat

The Problem With Ballast Water

What is ballast water? Ballast water is carried by ships in special tanks to provide stability and to optimize steering and propulsion. The National Oceanographic and Atmospheric Administration has calculated that 40,000 gallons (150,000 liters) of foreign ballast water are dumped into U.S. harbors each *minute*.

The problem with ballast water is very simply stated. Ballast water is taken up by a ship in ports and other coastal regions whose waters may be usually rich in planktonic organisms. It may be released at sea, in a lake or a river, in port, or in the open ocean along coastlines—wherever the ship reaches its destination. As a result, a myriad of organisms is transported and released around the world. Here are two examples.

Scientists studying an Oregon bay counted 367 types of organisms released from ballast water of ships arriving from Japan over a 4-hour period! Another study documented a total of 103 aquatic species introduced to or within the United States by ballast water and/or other mechanisms, including 74 foreign species.

The Veined Rapa Whelk *(Rapana Venosa)*

Rapana venosa (Figure I 22-2) is a predatory gastropod. Faculty at the Virginia Institute of Marine Sciences (VIMS) have been studying the whelk's impact on Chesapeake Bay. The following is a summary of their research.

Discovery of the whelk was purely by accident. A routine trawl in the lower reaches of Chesapeake Bay turned up an unknown organism, which was ultimately identified

FIGURE I 22-2 A veined rapa whelk, *Rapana venosa*. Note thickened edge of shell.

as *R. venosa*. This creature had left a trail of destruction behind itself in its wanderings, most recently decimating an oyster population in the Black Sea.

Excited by their discovery, the Chesapeake Bay researchers conducted another trawl, which turned up no individuals, but did collect two live masses of *R. venosa* eggs, which they returned to the laboratory and set about to hatch in a script beginning to take on similarities to the film *Aliens*.

As the eggs began to hatch, the scientists were eager to determine the tolerance of the hatchlings to variations in temperature and water salinity.

QUESTION 22-1. Why do you think the researchers at VIMS were interested in these data?

The scope of potential contamination of Chesapeake Bay by *R. venosa* and other introduced species is vast, considering that about 15 million tons volume of ballast water is dumped into Chesapeake Bay ports during a representative year solely by ships from ports with active *R. venosa* populations!

By careful study, the VIMS researchers learned something about the whelks' habitat preferences and diet. The whelks preferred hard sandy bottom into which they would quickly burrow: A 6-inch whelk could completely hide itself in less than an hour, leaving only its purplish colored siphon exposed, which it would instantly withdraw if disturbed. *R. venosa* spends at least 95% of its life buried, but it can and does move while burrowed, at speeds up to one body length per minute. They also learned that the whelk can feed and mate while completely buried.

QUESTION 22-2. How could knowledge of the whelk's preferred habitat be used by scientists to determine its potential range in an estuary like Chesapeake Bay?

The preferred diet of *R. venosa* is hard clams, but it will eat oysters, soft clams, or mussels if its favorite food is unavailable. Unfortunately for the clams, they share the whelk's habitat, and finally, there is a "healthy" hard clam commercial fishery in Chesapeake Bay.

The researchers were also interested in the whelk's predators, if any, in the organism's home waters. It turns out there are few: Octopi eat the whelks in the waters of southern Russia and the Black Sea, but there are no octopi in Chesapeake Bay. Other native whelks in Chesapeake Bay prey on smaller individuals of *R. venosa*, but there is an interesting twist: *R. venosa*'s shell is much thicker and more robust than native whelks (see Figure I 22-2). Moreover, *R. venosa*'s boxy shape means that adult rapa whelks are hard for other whelks to eat. Thus, if the rapana whelks can survive to adulthood, they have little to fear from the natives in Chesapeake Bay.

The researchers next turned their attention to what could prove to be *R. venosa*'s "Achilles heel," the egg and juvenile stages. They concluded that as eggs the whelks may be most vulnerable, as migrating fish could eat the bright yellow egg cases, or dislodge them, causing damage and perhaps death to the developing eggs.

QUESTION 22-3. Design a research project to study and measure the evolving impact of *R. venosa* on Chesapeake Bay, as well as its potential impact on other habitats along the East Coast. Suggest means by which the whelk's impact might be reduced or controlled.

There are numerous other examples of introduced marine species, including the lionfish (*Pterois volitans*), native to the subtropical and tropical Indo–Pacific Ocean, which was introduced in the Atlantic Ocean in 1992. Currently scuba divers report that they are fairly common along natural and artificial reefs along the southeast coast of the United States and in the Bahamas. The lionfish introduction differs from that of the rapa whelk in that the lionfish has venomous spines and may impact native commercial fish populations. For information on this and other invasive species, go to the U.S. Department of Agriculture National Invasive Species Information Center (www.invasivespeciesinfo.gov).

Media Analysis 1

Go to www.npr.org, and search for "Scientists Prowl to Destroy Mute Swan Eggs."[1] Listen to the 4:30 minute audio, and then answer the following questions.

QUESTION 22-4. Explain how mute swans got to Chesapeake Bay.

QUESTION 22-5. How do biologists kill (addle) mute swan eggs?

QUESTION 22-6. Why don't biologists just crush the eggs?

---

1. Or go directly to http://www.npr.org/templates/story/story.php?storyId=9923886.

QUESTION 22-7. In terms of numbers, how had the original 4,000 mute swan "infestation" been reduced by April 2004 when the audio was made?

QUESTION 22-8. What reasons did biologists give for advocating removal of the swans?

· ·

Media Analysis 2

Go to www.npr.org, and search for "Asian Oysters in the Chesapeake."[2] Listen to the 4 minute and 52 second report then answer the following questions.

QUESTION 22-9. On a sheet of graph paper, or on our website (http://www.jbpub .com/abel/), graph the decline in the Chesapeake Bay oyster harvest shown in Figure I 22-3 from 1885 to 2006.

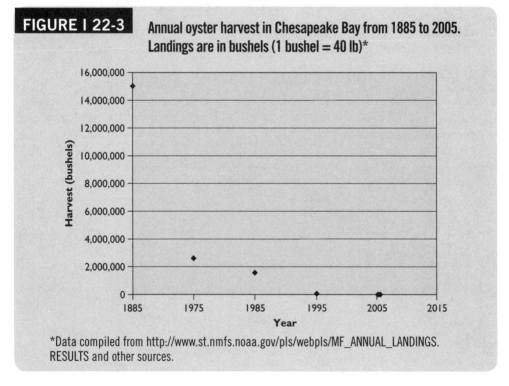

FIGURE I 22-3   Annual oyster harvest in Chesapeake Bay from 1885 to 2005. Landings are in bushels (1 bushel = 40 lb)*

*Data compiled from http://www.st.nmfs.noaa.gov/pls/webpls/MF_ANNUAL_LANDINGS. RESULTS and other sources.

QUESTION 22-10. What reasons did the report give for the decline of native Chesapeake Bay oysters?

QUESTION 22-11. One of the major causes of the oyster disease is nutrient pollution associated with human population growth in the region. Did the report mention this fact? Explain whether the report should note this fact.

---

2. Or go directly to http://www.npr.org/templates/story/story.php?storyId=1471571.

QUESTION 22-12. Commercial oyster fisher Jeff Hammer is described as being in the "vanguard of an effort to restore the oyster industry." Does the report describe anyone as being in the vanguard of an effort to restore Chesapeake Bay as an alternative to the introduction of Asian oysters?

QUESTION 22-13. Explain whether restoring Chesapeake Bay is an alternative to the introduction of Asian oysters.

QUESTION 22-14. What are the risks of Asian oyster farming or introduction in Chesapeake Bay?

QUESTION 22-15. Explain whether these risks are acceptable.

QUESTION 22-16. One citizen stated, "No matter how much study goes into this, I don't believe we can be 100% sure that we will not introduce another problem non-native species." Explain whether you agree or disagree with this statement.

QUESTION 22-17. Who are the stakeholders (interested or affected parties) in this issue? Explain which stake holder is most important and which is least.

QUESTION 22-18. Discuss alternatives to introducing oysters into Chesapeake Bay.

QUESTION 22-19. Discuss whether you would favor the introduction of Asian oysters into Chesapeake Bay.

Caulerpa in the Mediterranean

What is *Caulerpa?*

How is *Caulerpa* spread?

What are specific impacts of *Caulerpa* infestation?

KEY QUESTIONS

In 1984, French divers noticed a beautiful feathery emerald-green plant growing on about 1 m² of the sea floor off Monaco, adjacent to the Oceanographic Museum. Since then, *Caulerpa taxifolia* (Figure I 23-1) has spread along the Mediterranean coast and has altered and displaced native plant and animal communities. Early eradication was not attempted in the Mediterranean, and the infestation is now beyond control. As of 2004, *C. taxifolia* had infested over 30,000 acres of seafloor off Spain, France, Italy, Croatia and Tunisia.

QUESTION 23-1. Convert 30,000 acres to square meters (1 acre = 4047 m²). How many square meters were covered by *Caulerpa* in 2004?

Caulerpa is a single-celled, green alga with many nuclei. It may be the world's largest single-celled organism! Once introduced, *C. taxifolia* can spread from fragments and, in one minute, a broken-off fragment can form a new plant. *C. taxifolia* does not float, has never been observed to grow on boat hulls, and is unlikely to be transported in ballast water, but it can readily be transported on boat anchors and fishing gear.

Most *Caulerpa* species evolved in tropical waters, where native herbivores have immunity to toxic compounds within the alga. These tropical herbivores crop the algae and keep it from spreading. Moreover, *Caulerpa* is a prized food in many parts of the tropics.

Caulerpa species capable of surviving in temperate waters have few if any local predators as temperate water herbivores have no natural immunity to these toxins, allowing *Caulerpa* to grow unchecked if introduced to temperate waters.

 FIGURE I 23-1 *Caulerpa taxifolia,* perhaps the world's largest single-celled organism.

QUESTION 23-2. Why has *Caulerpa* spread so rapidly in temperate waters?

C. taxifolia can form a dense carpet on any surface including rock, sand, and mud. It is capable of growing up to one half inch per day (1 cm/day). *C. taxifolia* can grow in shallow coastal lagoons as well as in deeper ocean waters, possibly to depths of greater than 150 feet (nearly 50 meters).

Small colonies of *Caulerpa* have been found in a few places in coastal California, apparently from plants sold by aquarium shops. Assembly Bill 1334, signed into law in September 2001, prohibited the possession, sale, and transport of *C. taxifolia* throughout California. This bill also established the same restrictions on several other species of *Caulerpa* that are similar in appearance to *C. taxifolia* and that are believed to have the ability to become invasive. Earlier in 2001, the city of San Diego adopted an ordinance banning the possession, sale, and transport of any species of *Caulerpa* within city limits. Furthermore, the importation, interstate sale (including Internet sale), and transport of the Mediterranean strain (i.e., aquarium strain) of *C. taxifolia* is prohibited under the federal Noxious Weed Act (1999) and the federal Plant Protection Act (2000).

The extreme growth rate of *Caulerpa,* along with its ability to coat virtually any surface, has forced native plants out of large areas of the Mediterranean, with the effect of sharply reducing habitat available to native invertebrates and fish larvae.

QUESTION 23-3. What effect might the spread of *Caulerpa* have on Mediterranean fish stocks?

QUESTION 23-4. Should *Caulerpa* infestations in temperate waters be controlled by importing herbivores from the tropics? List advantages and disadvantages to such a strategy, if feasible.

Media Analysis

Go to www.npr.org, and search for "An Offshore Dilemma: Algae-Battling Tarps Now Habitats."[1] Listen to the 4-minute program about a *Caulerpa* eradication plan in California, and answer these questions.

QUESTION 23-5. How extensive an area was covered by the underwater tarp?

QUESTION 23-6. As of 2004, what area, in acres, of the Mediterranean was infested with *Caulerpa*?

QUESTION 23-7. How did the state of California kill the invasive seaweed?

QUESTION 23-8. What dilemma did state officials face arising out of their successful eradication program?

---

1. Or go directly to www.npr.org/templates/story/story.php?storyId=4231333.

PART 10

Energy from the Ocean

Renewable Energy From the Sea: Is It Limitless?

What is the potential for renewable energy from the sea?

What are the costs and benefits?

Is it cost-effective?

Who are the world leaders in the use of renewable energy from the sea?

What is slowing development of renewable energy from the sea in the United States?

With the dizzying ascent of oil, gasoline, and diesel prices during 2007 and 2008, Americans finally began to realize that the era of cheap fuels was coming to an end. As developing economies like India and China continue to industrialize and export more goods, thereby earning foreign exchange, they will continue to demand increasingly scarce transport fuels, whose use is still subsidized by governments in many countries, including India and China. Poor mileage among passenger vehicles and commercial trucks in the United States can only exacerbate the gap between an increasingly fixed supply and a growing demand for fuel.

Coupled with this trend, the world is beginning to grasp the ominous menace of rapidly accelerating climate change, with its potential to flood coastal communities and generate untold numbers of environmental refugees. In this Issue, we investigate the potential for renewable energy from the sea to supplant and eventually replace much of our reliance on fossil fuels and to assist in the battle against the worst effects of climate change. It would be ironic indeed if the sea, which poses the greatest risk of disaster from the impacts of climate change, could also be the instrument of humanity's salvation.

Energy From the Sea

In this Issue, we do not discuss offshore oil and gas deposits, as they are not renewable. We limit the topics to the following:

1. Offshore wind "farms"
2. Extracting energy from waves
3. Extracting energy from tides
4. Ocean thermal systems, which extract energy from heat differences between layers in the ocean (recall the concept of the thermocline from Chapter 6)

■ Offshore Wind Farms

Wind energy development is accelerating across the globe (Figure I 24-1). As of 2002, total installed capacity was about 40,000 MW. By the end of 2008, total installed capacity was expected to be 100,000 MW. Vast potential exists to extract energy from offshore winds and transport the electricity produced via cable to land. One advantage is that offshore turbines can be more powerful than those on land because of the more persistent and higher wind speeds offshore, and noise concerns are likewise minimal for sites farther offshore. Countries of the European Union already plan to install offshore wind farms in British waters and offshore Spain, Denmark, Germany, and Norway. The EU has set a goal of 20% energy from renewable sources by 2020. In its "Energi 21 Plan," the Government of Denmark set a goal in 1996 of providing 40% of the nation's electricity from offshore wind farms by 2030.

FIGURE I 24-1 Offshore wind farm. Describe advantages and disadvantages of siting large-scale wind "farms" offshore.

At the end of 2007, the British Government announced plans to vastly increase construction of wind power systems offshore, avoiding only sites of historic and environmental interest and shipping lanes. The goal was to have wind energy provide half of Britain's electricity by 2020. The North Sea is the most favorable site for development, as it is relatively shallow, has persistent winds, and is close to large centers of population.

In the United States, a research team from Stanford University concluded at the end of 2007 that a wind farm off California's Cape Mendocino could provide 5% of California's electricity. Most of the California coast, however, was deemed either too deep or too calm in the summer for widespread installation of wind farms.

Although winds are stronger and more persistent offshore, turbine installation and operating costs increase rapidly as the water depth and distance from shore increase. Thus, most installations are expected to be within 10 to 20 km from shore.

Concerns about large-scale installation of offshore wind farms center on their potential to disrupt shipping, their impact on the sea bottom, the effect of noise on marine life during construction, the effects of spinning turbines on wildlife, and the visual impact of turbines as seen from land. Research to date has shown that the turbines can be kept out of shipping lanes. Their impact on seabirds is minimal, and their visual impact depends on personal preferences. For example, even though a proposed installation of wind turbines offshore Cape Cod was opposed by a vocal minority of mostly wealthy property owners, surveys indicated that nearly 60% of residents favored the development.

Research on the impact of noise during construction on marine life is carried out at each proposed site.

■ Wave and Current Power

There is enormous energy in ocean waves, which are set in motion by the wind. Areas with the greatest potential lie between 30° and 60° latitude, along the western edge of South America, and the southern coast of Australia. Other favorable areas include the persistent trade winds belt north and south of the equator, and areas of high latitude (e.g., Iceland) characterized by high-energy polar storms. Wave energy alone could provide as much as 10% of global energy demand.

Total energy available depends on wave height, wave speed, wavelength, and the persistence of the waves. Using a variety of simple piston-like devices (Figure I 24-2), wave systems produce energy directly from the surface motion of ocean waves or from subsurface pressure variations. Although wave energy development is largely in the experimental stage, several nations have announced plans to encourage its development. The U.S. Department of Energy is funding research on wave power systems. Experts at the Electric Power Research Institute estimate that wave systems could produce 6% to 10% of the present U.S. demand for electricity. Wave power systems could, however, be most significant for islands like Hawaii, with no indigenous sources of fossil energy, high energy production and transport costs, and persistent waves.

A small wave power array is presently generating electricity off the coast of Portugal, and Pacific Gas and Electric, a California utility, has signed an agreement to purchase

FIGURE I 24-2 Concept of a wave energy generating system. Where are coastal sites in North America that are especially suited to wave energy development?

electricity from a wave array to be constructed off California's Humboldt County, beginning in 2010. In Oregon, a wave power device was successfully deployed in 2007 but sank when technicians tried to recover it after the test was complete.

Ocean currents also represent an enormous storehouse of energy (Figure I 24-3). Development of this promising source, however, is at an early stage.

■ Tidal Power

According to the U.S. Department of Energy, for tidal differences to be harnessed into electricity using present technology, the difference between high and low tides must be at least 5 meters (16 feet). There are only about 40 sites on the Earth with tidal ranges this favorable. The greatest is in the Bay of Fundy between the Canadian Provinces of Nova Scotia and New Brunswick, with maximum tidal range of 14 meters!

Although tidal power systems cost little to operate, construction costs and the potential for environmental impact are high, as a tidal power plant built across an estuary or bay would interfere with fish migration and could alter water characteristics such as dissolved oxygen and temperature. Other than tidal barriers, arrays of underwater turbines could produce electricity from tides, but could harm marine life caught in the turbines.

Tidal power systems have been successfully producing electricity since 1966 at a favorable site along the French coast of Brittany (Figure I 24-4). During the 30-year

 FIGURE I 24-3 The energy of ocean currents may be harvested in the future with underwater turbines. Given the distribution of the population of the U.S., which ocean current or currents would be most favorable for development?

FIGURE I 24-4 Tidal power generating station, La Rance estuary, France.

period from 1966 to 1996, the La Rance tidal power station, a 330-m dam across an estuary fitted with 24 10-MW turbines, produced 16 billion kWh of electricity without a mechanical breakdown.

A small (40 kW) tidal power station was installed in 2005 off the southeast Chinese coast.

Favorable conditions for tidal power generation exist in both the Pacific Northwest, between Washington State and the Aleutian Islands of Alaska, and the aforementioned Bay of Fundy.

■ Ocean Thermal Systems

The National Renewable Energy Laboratory states it eloquently:

> *On an average day, 60 million square kilometers (23 million square miles) of tropical seas absorb an amount of solar radiation equal in heat content to about 250 billion barrels of oil (the world's proven oil reserves total about 1,100 billion barrels). If less than one-tenth of one percent of this stored solar energy could be converted into electric power, it would supply more than 20 times the total amount of electricity consumed in the United States on any given day.*

This would be equal to about five times the global daily demand for energy.

Systems that convert solar energy stored in heated seawater to electricity are called Ocean Thermal Energy Conversion systems, or OTEC for short. They operate very much like home heat pumps: using warm seawater to vaporize a fluid like ammonia with a very low boiling point to turn a turbine hooked to an alternator to produce electricity. At least a 20°C temperature difference between warm and cold seawater is needed to efficiently produce electricity. Table I 24-1 shows typical temperature data from the tropical ocean off Tahiti.

QUESTION 24-1. How deep would a pipe have to be sunk to extract water of a sufficiently cold temperature, assuming that warm water was used from 20 meters?

TABLE I 24-1	**Temperature Profile, Pacific Ocean, Tahiti (ioc.unesco.org)**
Depth, meters	**T, °C**
20	27.2
100	25.3
200	21.4
400	11.1
600	6.5
1000	3.9
1200	3.3

Cold seawater in this case from a depth of 600 m or more would be used to condense the fluid. Warm surface seawater would vaporize it, and the process is repeated continually. The difference between power necessary to pump the seawater and the total power produced by the system is called the net output. A 1979 pilot project using a 50-kW OTEC system produced about 15 kW of net power. In 1981, Japanese scientists developed a 100-kW OTEC system that produced 31-kW net output.

Small-scale demonstration plants have been successfully tested; however, although OTEC systems provide enormous promise, research is still needed to scale up the process—for example, to produce pipe of sufficient diameter and resistance to corrosion and fouling to operate at depths exceeding 600 meters. Large-scale heat exchangers must also be made more efficient to produce more cost-effective electricity, and a commercial-scale plant producing 40- to 100-MW net output remains to be built and operated. By the middle 2000s, researchers felt that OTEC systems could be competitive with $60 per barrel oil and comparably priced natural gas.

QUESTION 24-2. What is today's price of oil on the world market?

Summary: Status of Ocean Energy Systems

Here is the late 2008 status of ocean energy systems.

- Offshore wind is being actively deployed and is expected to meet a significant proportion of Europe's electricity needs by 2020.

- Tidal power is commercial at a few sites around the world using present technology but faces significant environmental challenges.

- Wave power offers enormous promise in advantageous sites like the Pacific Northwest, parts of Western Europe, the southern Australian coast, and the southern coast of Chile. It is also feasible today for island nations with high fuel costs.

- OTEC systems could compete with $60 to $80 per barrel oil, especially in the tropics with a steep thermocline.

Media Analysis 1

We may not have to actually produce electricity from the energy of the oceans to help combat climate change. Go to www.npr.org, and search for "Looking to Oceans to Capture Carbon Dioxide."[1] Listen to the first 8 minutes of the 12:29 minute program. Then answer the following questions.

QUESTION 24-3. What is the Gaia hypothesis?

QUESTION 24-4. How does the Gaia hypothesis differ from a scientific theory?

---

1. Or go directly to www.npr.org/templates/story/story.php?storyId=14799203.

QUESTION 24-5. Where is Dr. Lovelock employed? What is his specialty, and where did his article appear?

QUESTION 24-6. What was Lovelock's target date for seeing large-scale human migrations to escape the hotter temperatures caused by climate change if nothing is done?

QUESTION 24-7. What does Dr. Lovelock refer to as the climate change "canary in the coal mine?"

QUESTION 24-8. Why does Dr. Lovelock propose that deeper ocean water needs to be "stirred up" to address climate change?

QUESTION 24-9. What is one objection to Dr. Lovelock's proposal, involving carbon dioxide in deeper ocean water?

QUESTION 24-10. What did Dr. Lovelock mean when he said his proposal would "use the Earth to cure its own problem?"

QUESTION 24-11. Why did Dr. Lovelock propose to test his idea near some coral reefs threatened by bleaching?

Media Analysis 2

Go to www.npr.org, and search for "Tidal Power: Tapping Energy From the Ocean."[2] Listen to the first 12 minutes and 30 seconds of the program. Then answer the following questions.

QUESTION 24-12. How did the speakers describe the potential off the U.S. coast for ocean energy?

QUESTION 24-13. How did Dr. Taylor describe the principle of using wave energy to produce electricity?

QUESTION 24-14. How much electricity does each buoy produce?

QUESTION 24-15. How did Dr. Taylor describe the advantages of wave energy?

QUESTION 24-16. What was the cost of electricity Dr. Taylor was aiming for, and when did he think his company would achieve it?

QUESTION 24-17. How could electricity produced by wave energy contribute to marine transport?

QUESTION 24-18. Why are sophisticated computers needed to convert the wave energy to electricity?

QUESTION 24-19. Where is the first tidal power project in the United States located?

2. Or go directly to www.npr.org/templates/story/story.php?storyId=12680697.

Media Analysis 3

Go to www.npr.org, and search for "Farming the Wind on the Gulf Coast of Texas."[3] Listen to the 4-minute program, and then answer the following questions.

QUESTION 24-20. What was Texas' rank as of 2005 in total wind energy produced?

QUESTION 24-21. How would wind turbines in the proposed array react to hurricanes?

QUESTION 24-22. How far does Texas' control extend offshore?

QUESTION 24-23. What organization won the 2005 Leadership Award for Renewable Energy?

QUESTION 24-24. How many wind turbines were proposed for the "wind farm" off Galveston, and how many homes would they serve?

QUESTION 24-25. There are about 100,000,000 "households" in the United States. How many similar arrays would be needed to provide all homes renewable wind energy (assuming that the wind blew all the time, or the electricity could be stored)?

---

3. Or go directly to www.npr.org/templates/story/story.php?storyId=4977598.

Methane Hydrates—Energy Boom or Climate Bust?

What are methane hydrates, and how do they form?

What geohazards might they generate?

Why are they formed where they are?

What role might they play as a future energy supply?

What role could they play in global climate change?

Hydrates are crystal structures made of water ice that trap natural gas (which is mainly methane) under conditions of high pressure and low temperatures. Hydrates could contain a vast trove of methane that potentially dwarfs other gas resources. On Alaska's North Slope alone, geologists estimate the hydrate potential at nearly 600 trillion cubic feet of methane—almost three times the known U.S. reserves.

Methane (CH_4) hydrates are complex structures formed when molecules of one type are trapped inside a lattice, or cage structure, formed by another molecule. In this case, the lattice-forming structures are water ice molecules, and methane molecules (among other things) are trapped inside. Such structures are called *clathrates* (Figure I 25-1).

Various types of clathrate structures may form all based on ice, depending on environmental conditions. More complex clathrates can contain larger molecules than methane, such as propane and isopentane.

There are two areas on Earth where methane hydrates are stored in vast quantities. One is in permafrost (permanently or seasonally-frozen ground in polar regions), and the other is in marine sediment within a critical range of temperature/pressure environments (Figure I 25-2).

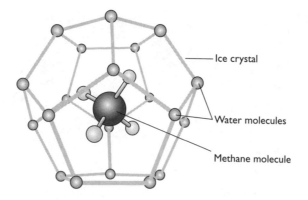

FROZEN GAS HYDRATE MOLECULE

★ **FIGURE I 25-1** Methane hydrate clathrate showing the methane molecule trapped within the ice crystal lattice. [*Source:* Adapted from Suess, E., et al., *Sci. Am.* 281 (1999): 76-83.]

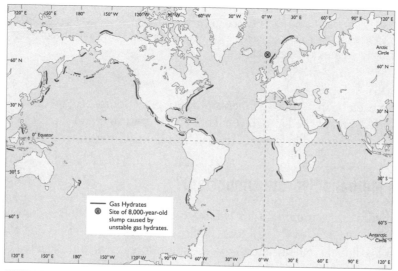

GENERAL DISTRIBUTION OF GAS HYDRATES

★ **FIGURE I 25-2** Map showing the locations of gas hydrate accumulations. Question: Briefly describe the global distribution of gas hydrates. [*Source:* Adapted from Suess, E., et al., *Sci. Am.* 281 (1999): 76-83.]

Methane Hydrates in the Gulf of Mexico

The United States Geological Survey considers the Gulf of Mexico to be the best setting on Earth to study gas hydrates. Here, they form mounds on the seafloor and can be readily sampled in sediments. Figure I 25-3 shows a hydrate mound on the sea floor. Hydrate mounds are often colonized by a variety of organisms. Bacteria "feed" on the hydrocarbons seeping out of the mound and in turn form the base of the local food web.

FIGURE I 25-3 A methane hydrate mound on the sea floor. How might a rise in sea level caused by global warming lead to an increase in hydrate deposits? What effect could this have in turn on greenhouse gases in the atmosphere?

Certain bacteria in marine sediments produce great volumes of methane when they feed on plant debris washed into the Gulf of Mexico from land and shore. Such biogenic methane is often trapped in layers of hydrate just below the seafloor surface.

Why Are Methane Hydrates Important?

Methane hydrates are important for two reasons. First, they may be a significant new source of a relatively "clean" fossil fuel (Figure I 25-4). Natural gas is one of our most important fossil fuels, but as natural gas consumption grows in the United States and globally, many experts worry that demand may outpace supply. We have already seen price spikes in natural gas in recent years, especially in 2008.

Although estimates vary, geologists believe that the resource potential of methane in gas hydrate exceeds the worldwide reserves of conventional oil and gas deposits, plus coal and oil shale, by a considerable degree; however, there are major obstacles to producing hydrates. For example, during transport to the surface (a low-pressure environment), marine hydrate decomposes and releases its hydrocarbons.

Second, methane is an enormously powerful greenhouse gas. It is much more effective at warming the atmosphere than the most common greenhouse gas, carbon dioxide (CO_2). Thus, any activity that releases significant new quantities of methane into the environment could accelerate global warming. A rapid increase in polar and presumably global temperatures 55 million years ago, inferred from sediment cores, could have been the result of large-scale methane release.

FIGURE I 25-4 Photo showing how easily gas hydrates burn.

In the following Media Analysis, we analyze the potential of using methane hydrates as an energy source and also assess possible environmental impacts.

Media Analysis

Go to www.npr.org, and search for "New Fuel Cell/ Methane Hydrates."[1] Listen to the last 38 minutes of the report (start at the 9-minute mark), and then answer the following questions.

QUESTION 25-1. Compared with other fossil fuels, about how much energy could be obtained from methane hydrates?

QUESTION 25-2. The Gulf of Mexico is described as an "unusual setting." In what ways is it unusual?

QUESTION 25-3. What did the methane hydrates look like from a submersible?

QUESTION 25-4. Natural gas or methane is repeatedly referred to as the "cleanest fossil fuel." What do you think this means?

QUESTION 25-5. What did the interviewer mean when he asked if mining methane hydrates was "opening a Pandora's Box"?

QUESTION 25-6. Describe how the horse mussel (*Modiolus*) uses methane as its own energy source.

---

1. Or go directly to http://www.npr.org/templates/story/story.php?storyId=1071693.

QUESTION 25-7. How does the methane in the methane hydrates form?

QUESTION 25-8. Caller Alfredo was concerned with the lack of focus on population control and energy conservation. How were his concerns addressed? Were you satisfied with the answers? Why or why not?

QUESTION 25-9. Would you describe the panelists as "cheerleaders," scientists, or skeptics? Support your answer.

QUESTION 25-10. How many times in the discussion was the environmental impact on the sea floor from removing the methane hydrates mentioned?

Basic Population Math

Completing the issues in this book sometimes requires knowledge of arithmetic and, in a few instances, basic math. We first provide a brief review of the metric system and scientific notation. These are followed by examples of how to use the compound growth equation.

The Metric System

The metric system is both simple and elegant. It is based on powers of ten: 10 millimeters in 1 centimeter, 100 centimeters in 1 meter, and so forth. This is a key point and should be kept in mind.

Throughout the text are questions that ask you make conversions from the English system to the metric system and back. Some common conversions as well as metric prefixes are given in Tables A 1–4. These may also be found inside the front cover of this book.

Here is a useful shortcut. To convert areas, you can use the conversion factors for the units of length and square them. For example, to convert 9 square feet to square yards, do the following:

$$9 \text{ ft}^2 \times (1 \text{ yd}/3 \text{ ft})^2 = 9 \text{ ft}^2 \times 1 \text{ yd}^2/9 \text{ ft}^2 = 1 \text{ yd}^2$$

Similarly, to convert volumes, use the factors of the units of length, and cube them.

Now answer the following questions as a test your ability to manipulate and convert these units. These skills are essential to answer the questions and understand the concepts in the issues in this book. Remember to use your conversion factors (so that units cancel each other out). For example, to determine how many liters are in 4.2 cubic meters:

$$4.2 \text{ m}^3 \times 1000 \text{L/m}^3 \times 4200 \text{ L}$$

QUESTION A-1. How many millimeters are in a meter?

QUESTION A-2. How many centimeters are in a kilometer?

QUESTION A-3. How many milligrams are in a tonne? (Tonne is the correct spelling for the metric unit of 1000 kg.)

QUESTION A-4. Express your height in feet, meters, centimeters, and millimeters.

QUESTION A-5. Express your weight in grams, kilograms, and pounds.

. . . .

Scientific Notation

Numbers that are very large or very small are most conveniently added, subtracted, multiplied, and divided by expressing the numbers as powers of ten.

One basic fact you need to know is that 10^0 (pronounced "ten to the zero, or ten to the zero power") is defined as 1.

Fortunately, converting large numbers to scientific notation and then manipulating these numbers are not difficult skills. First, convert large or small numbers to scientific notation, a skill with which you should already be familiar. Here is an example. In scientific notation, 23,000,000 is 2.3×10^7.

QUESTION A-6. Express 1 trillion (1,000,000,000,000) in scientific notation.

QUESTION A-7. Express 3,270,000 in scientific notation.

You can express the same number in a variety of ways using exponents, such as 32.7×10^5, 327×10^4, and so forth, and they all will mean the same thing. It is customary, however, to express all values in the same format, by placing only one digit to the left of the decimal place (e.g., 3.27×10^6).

QUESTION A-8. Express 23,000,000,000,000 (23 trillion) the customary way using exponents.

TABLE A-1	Metric Prefixes and Equivalents	
Large Numbers		
One thousand	=1000	=10^3 (kilo or k)
One million	=1,000,000	=10^6 (mega or M)
One billion	=1,000,000,000	=10^9
One trillion	=1,000,000,000,000	=10^{12}
One quadrillion	=1,000,000,000,000,000	=10^{15} (commonly used in expressions of energy use)
Small Numbers		
One hundredth	=1/100	=10^{-2} (centi or c)
One thousandth	=1/1000	=10^{-3} (milli or m)
One millionth	=1/1,000,000	=10^{-6} (micro or mc or μ)
One billionth	=1/1,000,000,000	=10^{-9} (nano or n)

For numbers summarized in scientific notation, the prefix in front of the unit (e.g., kilo) denotes the magnitude of the unit (e.g., kilograms = units of 1000 g). You should memorize these equivalents (Table A-1).

QUESTION A-9. Convert 2.43 mm to (a) nm, (b) µm (micrometers), (c) cm, (d) m, and (e) km. Express your answers as decimals and in scientific notation.

Manipulating Numbers Expressed in Scientific Notation

To add, subtract, multiply, and divide large numbers using scientific notation, you need to remember a few basic rules and check your work carefully.

■ Multiplication Using Scientific Notation

To multiply numbers expressed in scientific notation, multiply the bases and add the exponents.

For example, to multiply:

$$(5 \times 10^4) \times (2 \times 10^3)$$

multiply the bases:

$$5 \times 2 = 10$$

and add the exponents:

$$4 + 3 = 7$$

The result is 10×10^7 or, using the appropriate convention, 1.0×10^8. (This is the same as 100×10^6 and any number of other variations as well.)

■ Division Using Scientific Notation

To divide numbers expressed in scientific notation, divide the number in the numerator by the number in the denominator and subtract the exponent of the denominator from the exponent of the numerator.

For example, to divide

$$(8.4 \times 10^6) \div (4.2 \times 10^3)$$

divide the numerator by the denominator

$$8.4 \div 4.2 = 2$$

and subtract the exponent of the denominator from the exponent of the numerator

$$6 - 3 = 3$$

Thus, the answer is 2×10^3.

QUESTION A-10. Perform the following manipulations:

$$(9.5 \times 10^{-5}) \times (4.6 \times 10^{-10}) =$$
$$(7.2 \times 10^{24}) \times (2.9 \times 10^{20}) =$$
$$(9.5 \times 10^{-5}) \div (4.6 \times 10^{-10}) =$$
$$(7.2 \times 10^{24}) \div (2.9 \times 10^{20}) =$$

■ Addition/Subtraction Using Scientific Notation

To add numbers expressed in scientific notation, you simply add the numbers (i.e., the bases) after converting both to the same exponent. To subtract numbers expressed in scientific notation, you simply subtract the numbers (the bases) after converting both to the same exponent. For example, to add 6 billion to 11 million:

$$6,000,000,000 + 11,000,000$$

or

$$(6 \times 10^9) + (11 \times 10^6)$$

Convert both numbers to the same exponent (it does not matter which, but it is usually easier to use the smaller).

$$6 \text{ billion} = 6000 \text{ million or } 6000 \times 10^6$$
$$11 \text{ million} = 11 \times 10^6$$

Therefore,

$$(6,000 \times 10^6) + (11 \times 10^6) = 6,011 \times 10^6$$

An equally correct answer would be 6.011×10^9, which is the customary way to express the answer.

How to Project Population Growth Using the Compound Growth Equation

You will need a calculator with an exponent key. The equation is this:

$$\text{future value} = \text{present value} \times (e)^{rt}$$

where e equals the constant 2.71828…, r equals the rate of increase (expressed as a decimal, e.g., 5% would be 0.05), and t is the number of years (or other unit, as long as it is the same as r) over which the growth is to be measured.

Replacing words with symbols, this equation becomes this:

$$N = N_0 \times (e)^{rt}$$

The variable N_0 represents the value of the quantity at time zero, that is, the starting point.

This equation is central to understanding exponential growth and is one of the few worth remembering. Using that it is not as intimidating as you may think.

Sample Growth Calculation

Let us first project, a future human population, using the same example as on page 93 but this time using the compound growth equation.

The mid-year 2006 world population was 6.52 billion and was growing at a rate of 1.14%. Project what the world population would be in 2020.

Here's how to do this calculation:

$$N = (6.52 \times 10^9) \times e^{(0.0114 \times 14)}$$
$$N = 7.65 \times 10^9$$

■ How to Do This Calculation on a Typical Calculator

On a typical, nongraphics calculator, keystrokes are (commas are for punctuation only):

Key in 0.0114 (the decimal equivalent of 1.14%), then $\times$ (multiply sign), then 16, then $=$. This gives you the exponent (the number to which e must be raised). Next hit the button labeled e^x. On most calculators, e^x is above a button having another label (frequently ln). If this is the case on your calculator, you must hit the key labeled 2nd or 2nd F first, followed by the key with e^x above it. Finally, some calculators require you to hit the $=$ key after the e^x key.

Next, key in $\times$ (multiply sign), followed by the present value (6.52×10^9; on most calculators, this is done by keying in 6.52, then hitting the button labeled EE or EXP, and then keying in 9. If the EE or *EXP* label is above the key, you must first hit the key labeled 2nd or 2nd F before punching EE or EXP.) Finally, hit the $=$ sign.

Other Uses of the Compound Growth Equation

The compound growth equation can also be rearranged. If you know the starting and ending population sizes over a given period, you can calculate the average growth rate over that period using this formula:

$$r = (1/t) \ln(N/N_0)$$

Here is the same example used on page 94.

Given a 1987 world population of 5 billion and a 1999 world population of 6.0 billion, calculate the average annual growth rate over the 12 year period.

$$r = (1/t) \ln(N/N_0)$$
$$r = (1/12) \ln(6 \times 10^9)/(5 \times 10^9)$$
$$= 0.0152, \text{ or } 1.52\%$$

Stop here and think about the size of the rate of growth between 1987 and 1999 that led to an increase of one billion people. Did the rate 1.52% seem large to you? Thus, you can see how relatively small rates of growth can lead to very large increases over short time intervals.

You can also calculate how long it would take a population of a given size to grow (or decrease) to a different size at a specified growth rate using

$$t = (1/r) \ln(N/N_0)$$

Here is the example used on page 95.

At an annual growth rate of 1.14%, how long would it take a population of 5 billion to grow to 6 billion?

$$t = (1/r) \ln(N/N_0)$$
$$= (1/0.0114) \ln(6 \times 10^9)/(5 \times 10^9)$$
$$= 15.99 \text{ years}$$

· ·

Doubling Time

For any population that is growing exponentially, the time it takes for the population to double can be estimated as follows:

$$t = 70/r$$

Where t = the doubling time and r = the growth rate, expressed as the decimal increase or decrease $\times$ 100 (you would enter 5 for 5%, or 0.05, increase).

For example, calculate the doubling time for a population growing at a rate of 7% per year.

$$t = 70/7 = 10 \text{ years}$$

This means that a population growing at a 7% annual rate will double in 10 years and will double again in another 10 years. Thus, at the end of 14 years, the population will be four times as large as originally.

Another useful piece of information is that it takes 10 doubling for a population growing exponentially to grow by a factor of approximately 1000 (actually, 1024).

Doubling time is a very useful and practical concept for projecting and analyzing the implications of growth. When a population is decreasing, you can use the same equation to calculate the halving time.

INDEX

CREDITS

Page 1 © Digital Vision; **Page 2, 7, 11, 14, 17** © AbleStock; **Page 42** Courtesy National Geophysical Data Center/NOAA; **Page 69** Courtesy of Monika Bright, University of Vienna/hydrothermalvent.com; **Page 72 (Part A)** © Jeanne Poindexter/Visuals Unlimited; **(Part B)** © T.E. Adams/Visuals Unlimited; **(Part C, Part D)** © David Phillips/Visuals Unlimited; **Page 74** © Phototake, Inc./Alamy Images; **Page 79** © Paul Yates/Shutterstock, Inc.; **Page 80 (B)** Courtesy of Eduardo A. Morales/Patrick Center for Environmental Research, The Academy of Natural Sciences of Philadelphia; **(C)** © John Cunningham/Visuals Unlimited; **Page 82 (A)** © Elisei Shafer/Dreamstime.com; **(B)** Courtesy of NOAA; **Page 86–87** © Photodisc; **Page 89** © David Davis/ShutterStock, Inc.; **Page 90** © Rick Dove; **Page 98** © Photodisc; **Page 102** Courtesy of Jeremy Weiss and Jonathan Overpeck, The University of Arizona; **Page 104** © Photodisc; **Page 107** Courtesy of FEMA; **Page 111** © Golf Money/ShutterStock, Inc.; **Page 112** © Jeff La Marca/Flickr.com; **Page 116–117, 126** Courtesy of SeaWiFS Project/NASA/GSFC/ORBIMAGE/NOAA; **Page 130** © Piccaya/Dreamstime.com; **Page 132** Courtesy of SeaWiFS Project/NASA/GSFC/ORBIMAGE/NOAA; **Page 133** Image by Donna Thomas/MODIS Ocean Group NASA/GSFC SST, product by R. Evans et al, University of Miami; **Page 136** Courtesy of SeaWiFS Project/NASA/GSFC/ORBIMAGE/NOAA; **Page 137** Courtesy of Debbie Larson/NOAA; **Page 142–143** © Mark Kuipers/ShutterStock, Inc.; **Page 147** © 2008 Erin Heydenreich, Center for Whale Research; **Page 150** © Mark Kuipers/ShutterStock, Inc.; **Page 151** © Myrleen Ferguson/PhotoEdit, Inc.; **Page 157** © Mark Kuipers/ShutterStock, Inc.; **Page 161** Courtesy of Jacques Descloitres, MODIS Land Rapid Response Team/NASA/GSFC; **Page 171** © Mark Kuipers/ShutterStock, Inc.; **Page 175–176** © AbleStock; **Page 177** © Marevision/age fotostock; **Page 179** © Jose Antonio Sanchez/ShutterStock, Inc.; **Page 183, 193** © AbleStock; **Page 194** © Mike Johnson/Seapics.com; **Page 195** Courtesy of Jamie Hall/NOAA; **Page 202** © AbleStock; **Page 207–208** Courtesy of Florida Keys National Marine Sanctuary/NOAA; **Page 209** © Photos.com; **Page 217** Courtesy of Florida Keys National Marine Sanctuary/NOAA; **Page 218** Courtesy of OAR/NURP/NOAA; **Page 226–227, 237** © AbleStock; **Page 238** © The Photolibrary Wales/Alamy Images; **Page 239** Courtesy of the Fisheries Collection/NOAA; **Page 240** Courtesy W. Perryman, National Marine Fisheries Service, NOAA; **Page 241** Courtesy of the Fisheries Collection/NOAA; **Page 248** © AbleStock; **Page 249** Courtesy of the Fisheries Collection/NOAA; **Page 251** Courtesy of William B. Folsom/NOAA Fisheries; **Page 253** © AbleStock; **Page 254** © Images & Stories/Alamy Images; **Page 255** © Wayne G. Lawler/Photo Researchers, Inc.; **Page 261–262** © Christopher Marin/ShutterStock, Inc.; **Page 263** Courtesy of the U.S. Geological Survey; **Page 265** Courtesy of the U.S. Geological Survey Archive, U.S. Geological Survey, Bugwood.org; **Page 269** © Christopher Marin/ShutterStock, Inc.; **Page 270** Courtesy for Alexandre

Meinesz, Universite de Nice Sophia-Antipolis; **Page 272–273** Courtesy of NOAA; **Page 274** © Dervaux/EWEA; **Page 276** Courtesy of Ocean Power Technologies, Inc.; **Page 277 (top)** Courtesy of Marine Current Turbines, Ltd.; **(bottom)** Courtesy of Nova Scotia Power, Inc.; **Page 282** Courtesy of NOAA; **Page 284** Courtesy of Ian R. MacDonald/Texas A&M University-Corpus Christi; **Page 285** Courtesy of the Naval Research Laboratory/NOAA; **Page 287** Courtesy of NOAA

Unless otherwise indicated, all photographs and illustrations are under copyright of Jones and Bartlett Publishers, LLC, or were provided by the authors.